EINSTEIN A~~ ~~~~ ~~ ~~~~
In Search
Of The Cosmic Man

by William Hermanns

BRANDEN BOOKS, Boston

Library of Congress Cataloging in Publication Data

Hermanns, William.

  Einstein and the poet.

Bibliography; p. 1 Einstein, Albert, 1879-1955. 2.
Science—Philosophy. 3, Religions (Proposed,
universal, etc.) 4. Physicists—Interview. I. Einstein,
Albert,
1879-1955. II. Title.
QC16.E5r1394   1983   :530'092'4 [B[83-3776
ISBN (1-8283-1851-4
ISBN 0.8283-1873-5 (pbk

E-Book Edition ISBN 9780828324113

Branden Books PO Box 812094 Wellesley MA 02482
www.brandenbooks.com

Dedication

To Elsa Brandstroem, to Raoul Wailenburg
and to the youth of the world:

You, Elsa, angel of the prisoners of war,
have saved hundreds of thousands of Germans in
Siberia from starvation and death in the First World
War.

You, Raoul, saved many thousands of Jews
marked for the Nazi gas chambers.
You both were Swedes.

What an example for you, youth of the world,
how unconditional love is stronger than the rules
of the State.
May this book help make you aware of your spiritual purpose.

Let Einstein show you the way.
Accept and believe him, just as he always believed in you.
Your foresight and resolution determines the fate of humanity,
whether it will be saved or doomed.

Acknowledgements

I wish to thank Kenneth Norton-Hermanns for encouraging me to publish the conversations and assisting me with typing. Through his companionship I gained a son, he a father, and the world my conversations with Einstein.

Contents

The First Conversation

Introduction

When in 1927 Goebbels became Gauleiter in Berlin, I was so shocked about his propaganda campaign for Hitler's book, *Mein Kampf,* reflecting the pathological madness of the coming Fuehrer, that I took a public stand. Preparing for a position in the League of Nations, I had sociology as my major field, and I saw in Hitler not so much as a personal demon, but as a reflection of the plague of hate in the slumbering mind of most German men and women which could break out and spread to world war. I attempted to arouse the Humboldt Club in Berlin, whose membership consisted mostly of students in the field of international relations. My efforts were in vain, however, for many wore swastikas, although still hidden, under their coat lapels. In spite of my defeat in the Humboldt Club, I was able to convince the members of the League for Human Rights —I was one of its Honorary Secretaries —to write a letter to the German President, von Hindenburg, depicting the danger of the coming dictatorship for Germany if not for the world. I warned him that Hitler's book would be able to change the mass mind of the German people to a revolutionary mind that would don the military garb and smash the Weimar Republic and its pillar, the President himself. We learned that Hindenburg's son, as well as Meissner, his Secretary of State, were encouraging compromise with Hitler's future plans, since Hindenburg's purchase of his new estate, Neudeck, was involved in the Osthilfe Scandal and susceptible to Hitler's political blackmail. After conversations with representatives of the Democratic Party in the Reichstag, especially with Gertrud Baeumer and Elizabeth Lueders, I became convinced that the only person in Germany who could be a forceful counterbalance to the man Hitler as well as to his strategy, already discernible in the marching feet of his Nazi organization, was Albert Einstein.

I had become acquainted with Einstein in 1921 in Berlin after I returned from forty months of French captivity and began to study international law and sociology for a diplomatic position in Geneva. I once even had the opportunity, following a welfare concert in Berlin where Einstein played the violin, to take part in a conversation he held with a group of students and to read to him my poem, *Verdun,* which prompted him to say to the some twenty listeners around him, "This poem should be known by our youth; then they would dislike the new uniforms."

VERDUN
At dusk an awesome shiver
runs through my ancient stone.
There draws around my walls
a pilgrimage of bones
They stop before my gate
and knock. I let them in.
They move to my cathedral
march with a ghostly din
in uniform of rags.
Some have no skulls or hips,
some chests still wear a medal
and dust falls from their ribs.
At dawn the visit ends.
My sons, my noble waste
of hopes, slip over the bridge
back to their holes in haste.

Now my cathedral trembles,
the windows clack and shiver;
they try to shrink, to hide.
The dawn shines off my river,
wipes darkness from my walls.
The light prepares a feast
for the hungry mouths of guns.
They gather in the east
I stand—I have no choice —

look to my towers and cry,
Til stretch you like two fingers
that swear unto the sky:
Oh Teuton fury listen,
the gates of hell you shake
but not a mother's heart.
I stand and shall not break!"

Oh river, do you hear
my vow? Come, cease to flow.
Rise up and cool my wounds!
The Crown Prince starts his show.
His iron vultures rise
on wings of thundergales.
They dive to tear my children
and scatter their entrails.
My houses tumble down;
Death spits her flame and smoke
upon the roofs and streets —
what is alive must choke.
As far as I look, the earth
heaves up in waves and bends.
This is the feast for rats —
the Blond Beast feeds his friends.

Oh Meuse, you feel my sorrow.
You see me drain my cup.
There! On your bridge they march —
my last sons I give up.
Oh river, rise and cool
my burning pain.   What fate:
Dawn sees my sons in battle,
dusk herds them to my gate.
I, mother of France, keep watch.
Man's hate has shackled me.
When will dawn bring one love

and one humanity?
There is a knock at my gate.
My youngest — no more eyes!
Oh river, cool my pain
and let your waters rise!

Einstein's remark encouraged me to distribute the poem in season and out of season, everywhere, to stir the German conscience. Moreover, I had introduced on the Berlin radio educational plays in which I tried to show the virtues of democratic freedom, and my plan to use Einstein as a counterweight against Hitler in a radio lecture was enthusiastically accepted by the Berlin Broadcasting System. By this time I saw Hitler and Einstein like two mighty lighthouses beaming opposing beacons throughout Germany. Goebbels was boasting that a million youths were ready to march whenever Hitler called them. I felt if one man in Germany could be the beacon of democratic principles to the youth, that man would be Albert Einstein, already marked by the Nazis as "Enemy Number One of the Nation," and the object of at least seven plots to take his life.

Overwhelmed with the conviction that my miraculous survival at the Battle of Verdun was for the purpose of serving humanity, 1 could not stand aside, an observer, as the Nazis led my country into the abyss. Using my success as a playwright, I resolved to broadcast my findings on the nature of true genius, a quality born in man and not made by others. Perhaps with Einstein I could Ibrm a new movement for youth. I was ready! I would redeem my vow at Verdun to serve God. I would help Einstein try to save Germany from the abyss Hitler was preparing. But how to approach him?

I wrote to him of my intention and soon received word that I should visit him the following Tuesday afternoon. I felt my time had come. A shadow from another world, however, an ominous omen, was to appear before I could set foot over the threshold of Einstein's apartment.

First Conversation with Einstein

March 3, 1930

It was a chilly day, and while climbing the stairs to Einstein's fourth-floor apartment on the quiet Haberlandstrasse, I wondered how to fit my prowess, or what Bergson called *élan vital,* into his world of shyness, since he usually restricted his social life to his wife and a handful of close friends. Above me, I suddenly saw Einstein rushing down from the attic, his hair flying about his head. No sooner had he reached the landing than a man emerged from one of the dark corners of the dimly lit staircase and jumped between him and the door to his apartment. The stranger must have heard me approaching, for he suddenly retreated. Einstein took advantage of the moment to unlock his apartment and disappear inside, slamming the door behind him. The stranger and I exchanged uneasy glances: I smiled uncomfortably and asked him if he had an appointment also.

"Yes," he shot back with a fearful expression.

"But that was Einstein whom we just saw," I replied.

"I know that," he nodded, "but you go in first. I shall see him later."

I kept glancing at him as I rang the bell. He was tall, with a thin almost starved appearance. He was about forty, was poorly but neatly dressed, and wore a black ribbon tied around his soft collar in an artist's bow. He seemed to gaze at me much as a small boy would after his mother had caught him pilfering the cookie jar. My eyes adjusted to the dimness and I noticed a small undernourished woman in a dark-blue tailored suit standing behind the man, her head covered with a weather-beaten black felt hat. Was this encounter, too, a meaningful coincidence?

A maid opened the door cautiously. "There are others waiting outside," I said as I entered.

They have been here for hours," she whispered. She took my overcoat, hat and gloves and led me into the drawing room. It was decorated in the popular Biedermeier style, with green wallpaper and ornamental vases. Near the window was a grand piano, on top of which were two violin cases. There were paintings and family portraits and a plaster bust of Goethe, and also some wax figures representing bearded and ascetic Biblical characters in colorful flowing robes.

As my eyes wandered around the room, my thoughts returned to the man in the hall. He reminded me of a day in 1921, when I had just returned from French captivity. I saw students celebrating the murder of Erzberger, marching around the court of the University of Berlin and singing:

Swastika on helmet,
Armband black-white-red,
Storm Detachment Hitler
Is our name.

These students were all lean and hungry-looking, shabbily dressed in partial uniforms; but the one who led them was adorned in elegant riding boots, an officer's tunic, and a medal: the Iron Cross. I couldn't help stopping him with the words, "What does that song mean?" Seeing how they closed in on me, I added, "I have the Iron Cross, also—I was at the Battle of Verdun." Then they stood more or less at attention and the leader said, "The song means we will change Germany. A few years more and then no more Jews. They have betrayed us in war with a stab in the back."

"I've heard Ludendorff say these very words," I responded, "but what you don't seem to know is that he had to drop this charge when the statistics of the army revealed that proportionally as many Jews were distinguished for bravery and commissioned as officers as were the Catholics and Protestants," They looked at me for a moment, turned around and went on singing.

Now here at Einstein's door, I met a man as lean and hungry looking as those students from years before who still haunted me. Was this an ill omen? I suppressed these recollections and continued to survey Einstein's home. Through a half-opened door I caught a glimpse of the living room, in the center of which a seated man was modeling Einstein's face for a bronze plaque. What alcove of what scientific institution, I wondered, would dare to hang this huge bas-relief? Probably none in Germany; for, who, in those perilous days, would have openly proclaimed those achievements? It had been commissioned, I supposed, by the British, for they were Einstein's greatest admirers.

Einstein entered, a bulky man of about fifty, carrying a pipe and wearing a simple grey coat. He was taller than I had thought. To conceal my somewhat awkward feelings, I asked questions about the delicate Biblical figurines. He showed me each one with pride and careful tenderness. "They were made by my daughter, Margot," he said, as he held them in his perfectly smooth hands.

As we sat down near the large windows, Einstein smiled, his eyes calm but curious as I took a notepad from my pocket. I looked out at the tops of the trees that grew along both sides of the street below, wondering how to break the silence. But I could feel his clear brown eyes penetrating me so guilelessly that, losing my restraint, I blurted out, "How did you become a genius?"

"Who says I am one?" he retorted briskly, throwing both hands forward, palms up.

Realizing that I should curb my enthusiasm, I apologized and asked him how he had discovered his theory of relativity.

"It is difficult to state just when it began," he replied with a smile.

Watching him as he filled his pipe, I was amazed by the graciousness with which he encouraged the curiosity of a stranger.

"When I was about five," he went on, "my father gave me a compass as a toy. The movement of the needle so intrigued me that I could hardly sleep. I wanted to find out why the needle never deviated, even when I turned the compass round and round. So I asked my father whether or not there was a pull from the outside which made the needle point consistently toward the north. He said there was such a pull, but couldn't tell me what caused it. When I asked my uncle, an engineer, he immediately proceeded to teach me some fundamentals of algebra, with this advice: 'What you don't know, call x, then hunt until you find what it is.' From that time on, I have called everything I didn't know x, especially magnetism."

While he talked he smiled the innocent smile of a child, as if the reminiscence rejuvenated him. His eyes sparkled and radiated warmth, the graying hair belied by the youthful expression.

"By the time I was twelve," he continued, "a thin schoolbook on geometry had become my most holy possession. The propositions about straight lines and triangles filled me with so much admiration that I vowed to think like Euclid, and soon began to look at everything around me with geometrical eyes."

He shrugged, seeming to speak more to his pipe than to me, repeating that his fascination was with Euclid, whose lucid and pure abstractions had shown how powerful an instrument the human mind could be. His classmates found Euclid puzzling; the ancient mathematician made no impression on them, but Einstein, the young student, did. In his consuming enthusiasm for this ancient Greek, he was considered very odd. He was utterly oblivious to the sports and games of his classmates, preferring music and geometry.

"Isn't loneliness the price the world exacts from a genius, and don't you feel like one?" I asked.

Einstein chuckled; then frowned, looking at me out of the corner of his eye. Finally, with a roguish twinkle: "Someone believed in me — my mother. When I was four or five years old, she took me to see my aunt. While we were at dinner, my mother said, 'Mark my words, Albert will become famous!' I never forgot how they all looked at her. My aunt spread this news among my relatives, and how my cousins made fun of me!"

"But wasn't your mother right?"

"Nonsense! Women's chatter."

Einstein drew deeply on his pipe and blew smoke into the air. "School failed me, and I failed the school. It bored me. The teachers behaved like *Feldwebel* (sergeants). I wanted to learn what I wanted to know, but they wanted me to learn for the exam. What I hated most was the competitive system there, and especially sports. Because of this, I wasn't worth anything, and several times they suggested I leave. This was a Catholic school in Munich. I felt that my thirst for knowledge was being strangled by my teachers; grades were their only measurement. How can a teacher understand youth with such a system?"

I ventured, "You were an early advocate for educational reform, weren't you?"

"I don't know what I advocated, but from the age of twelve I began to suspect authority and distrust teachers. I learned mostly at home, first from my uncle and then from a student who came to eat with us once a week. He would give me books on physics and astronomy.

The more I read, the more puzzled I was by the order of the universe and the disorder of the human mind, by the scientists who didn't agree on the how, the when, or the why of creation. Then one day this student brought me Kant's *Critique of Pure Reason.* Reading Kant, I began to suspect everything I was taught. I no longer believed in the known God of the Bible, but rather in the mysterious God expressed in nature."

"How amazing," I said, "that a poor student who happened to eat at your father's table should mold your thinking. Wasn't that providential?"

Einstein gazed at me askance; had I said something wrong? He blew out a dense cloud of smoke. "But then I found an idea that was to occupy my thinking for years."

"How old were you?"

"About sixteen. It was in a boarding house in Switzerland near where I went to school."

He closed his eyes and smiled to himself. In spite of the grayish tinge of his thick curly hair, for the moment Einstein resembled a sleeping boy. "Look," he seemed to say; "only a boy can be a thinker. He who considers himself grown-up and mature forfeits the wondering wisdom of a child."

Indeed, as he spoke again, his face still reflected the curiosity which had gripped him more than thirty years before. "Early one morning while I was sitting on my bed and dressing, I watched light travel through the window."

"Light travels? I never knew that," I interrupted.

Einstein's dark brown eyes opened wide. "What then is your theory of light?"

"Well, for me, light is when it is day, and dark is when it is night."

"Oh, come now, you are no longer in grammar school. Have you never had physics?" He sounded amused. "I expect you were ill at the time!"

I felt both stupid and anxious. Another comment like that, I thought, and he might dismiss me, with good reason. I muttered at the idea of light traveling had never crossed my mind, I was more interested in...(I nearly said poetry but quickly caught myself and mentioned two fields certain to be of interest to him:)...philosophy and sociology.

"But isn't philosophy closely connected with physics and mathematics? You must have realized that."

I shook my head, and there ensued an embarrassing silence. I hadn't visualized myself sitting here with my knowledge of physics under trial, and felt so ill-prepared that I wondered if Einstein was about to pull out his watch. But his voice put my worry to rest.

"Each light ray travels through space, and it occurred to me that I could measure the path of a ray from one point to another. I had already been taught that light passes through the ether, and I asked my teachers if it were possible to measure such a path and thus learn more about the relationship between light, the ether, and the rotation of the earth. Some didn't understand; others laughed or suggested I read certain books."

He told how these books bothered him. Whether concerning mathematics or natural science, their knowledge was subdivided, while he wanted universal principles. Even at the age of sixteen he felt that life was too short to be spent groping about in specialized fields. As he said this, he sat up straight in his chair and waved his pipe at me. The basic laws of the universe are simple, but because our senses are limited, we can't grasp them. There is a pattern in creation."

He pointed with his pipe to some scores on a chair beside the grand piano. "Every celestial body moves through space with the harmony of a violin moving through a Mozart symphony. Or," he asked with a quick turn of his head, "do you believe God is an enigma?"

His question startled me. "No," I hastily replied.

"Why, then, shouldn't it be possible," he went on, "to uncover an absolute truth? Euclid gives us confidence in the power of thought, and I have no doubt that it is possible for man to determine principles that would be as valid on the moon or the stars as they are on earth. If I stay, then, in my own little corner and observe the structure of the universe, what I find will be true for me and also for you. The criterion of truth is that it be true under all conditions — that it be the combination and sum of all relative observations. With this in mind, I decided to travel with a ray of light into space to find out what it is and what it does."

Einstein, seeing my puzzled expression, leaned over, brushed his hair back from his eyes, and explained, "You see, my young friend, if a train passed you and you wished to see who was inside, you would have a good chance of finding out if you were on a bicycle riding next to it. Your chances would be better still if you were in a fast car."

"But you just told me our senses are limited and we can't depend on our eyes to tell the truth."

He laughed like a schoolboy. "All! But we have higher mathematics, haven't we? This gives freedom from my senses. The language of mathematics is even more inborn and universal than the language of music; a mathematical formula is crystal clear and independent of all sense organs. I therefore built a mathematical laboratory, set myself in it as if I were sitting in a car, and moved along with a beam of light."

While talking he moved his hands gracefully, rhythmically. "I had always been fascinated by light. When I switched on a light bulb, light flooded the room, just as light from the sun Hoods the earth. I knew, too, that it travels at about 186,000 miles a second and that it takes eight minutes to reach the earth from the sun, which is about ninety-three million miles away. So you must understand that the sun rises each morning eight minutes before it seems to do so."

As he spoke, his hands, soft, fine and unwrinkled, fascinated me. Along with his brown eyes, which shone with endless wonder, they suggested eternal youth rather than a middle-aged man.

"Now they taught me in school," Einstein continued, "that the universe is filled with an ether which carries the waves of light, but since none of my teachers could tell me what this ether was, I had some thinking to do. 'Suppose,' I said to myself, 'there is no ether? Could light waves possibly travel without a medium? Or perhaps an ether does exist, but not in the way I've been taught — maybe it is fixed and not in motion,' And the Newtonian hypothesis that light travels in a straight line also bothered me."

At this point I was so intrigued that I cast away all thought of directing the conversation; my pencil trembled in my fingers as I wrote down his words. He told me that the celestial harmony, the only part of his religious instruction he had always accepted as true, was reaffirmed by his experiments with light. "There is a pattern in the celestial bodies that can be analyzed like a Bach fugue. The stars don't exert a direct physical force on each other from across millions of miles of empty space, but rather define and determine the properties of the space around them by their gravitational fields. These gravitational fields curve space into mountains and valleys, and the stars and planets follow the lines of least resistance, like marbles that a child shoots on the ground. The theory of relativity is thus as comprehensive as creation itself. All nature expresses mathematical simplicity; or to put it differently, without pure thought —in our case, mathematical construction —one cannot comprehend the phenomena of nature which constitute reality."

The word "simplicity startled me, and I asked what he meant. Einstein patiently explained that the physicists had clung to the concept of ether because, without it, they were unable to rationalize action at a distance. This hypothetical ether, however, had hindered rather than helped a unified view of the nature of forces.

"The ether had never been observed, so I discounted it. I

started all over again with a fresh view of nature. My desire, or shall I say my mission, has always been to simplify human life by simplifying human thought. My theory states that matter and energy are the same, and I was thus able to unite the field laws of mechanics. I was never overawed by cherished concepts in physics, no matter how time-honored. Newton declared that space and time were absolute, and I consider time a fourth dimension. Newton also held that gravitation was a force, and I consider it to be a curvature of space."

Einstein glowed like a primitive man who had just discovered fire by striking two stones together. "Mass is nothing but concentrated energy. So the sun, radium and uranium are no longer mysterious: we now understand how they are able to eject light or particles at enormous velocities and continue to do so for millions of years."

"All this," I ventured, "may seem simple to you, but many people are more confused than ever. According to your theory, if a twin travels on a train at the speed of light for one year, and then meets his brother, he can say, Tm a year younger than you.'"

Einstein smiled. "I have used this example myself, for it does not conflict with my theory—only with our notion of time."

"Couldn't you," I asked, "use your mathematical laboratory to move faster than the light beam?"

"That isn't possible, or I would see my own birth," he said matter-of-factly. "The light you see this evening may have started out from a distant star before Christ was born, yet it reaches your eyes only now. Likewise, what happened in Paris during the French Revolution could be seen today by someone on a distant star, if he had a sufficiently powerful telescope. On a star still farther away, people may see us sitting here a thousand years hence."

"What is the most sensational party of your discovery?" I asked, continuing my note-taking.

"My contemporaries' explanations of it. Since others have explained my theory, I can no longer understand it myself," Einstein mischievously replied.

"But do you know that Henri Poincare claims you are greater

than Copernicus?"

He wagged his head happily and chuckled, "Poincare is right. Not about me, alas, but in his own contributions to geometry." He then added that such comparisons were sometimes misleading. "Galileo was the first to submit geometrical theory to experimentation, and will forever remain the father of modem physics. And Newton's theory of gravitation was probably the human mind's greatest attempt to explain natural phenomena. But even he was restricted by the knowledge of his time. And I, too, cannot jump over my own shadow." There was a pause while he gazed at the ceiling. "As for Copernicus, both he and I have in common a deep dislike of dictatorship, which opposes free thinking."

The word "dictatorship" brought to mind recent events and the man outside Einstein's home. I said to Einstein, "We both are members of the League for Human Rights. Did you see the League's memorandum in which Hitler, on his arrival in Berlin, announced that 'from the trenches comes the pure man of iron-will to lead the Fatherland back to honor and glory'?"

Einstein shrugged his shoulders and mumbled, "It does not interest me. I am a socialist, as you know, and my only interest is to teach youth to consider the welfare of all people and to attain the intellectual freedom of the individual which is required to build a socialist state. I believe that if Hitler should come to power, it will mean the complete enslavement of the individual."

As Einstein again filled his pipe, I asked whether he was aware that his colleague, Professor Lenard, had stated that if Einstein's theory of relativity is correct, it is German science administered in a Jewish vessel.

His gentle, contemplative expression immediately turned hard and bitter. "This bears out what I have recently said: if my theory is right, the Germans will call me a good German and the French will call me a good European. And if it is wrong, the Germans will call me a Jew, and the French, a German. But I don't care whether they praise me or damn me. I am a link in a scientific chain; my students will carry on where I leave off."

"Do you believe," I asked, "as Max Weber once put it, that 'the difference between art and science is that art is always finished, science never'?"

He nodded, twisting little tufts of hair, "Science is never finished because the human mind only uses a small portion of its capacity, and man's exploration of his world is also limited." He pointed towards the window, "If we look at this tree outside whose roots search beneath the pavement for water, or a flower which sends its sweet smell to the pollinating bees, or even our own selves and the inner forces that drive us to act, we can see that we all dance to a mysterious tune, and the piper who plays this melody from an inscrutable distance — whatever name we give him — Creative Force, or God —escapes all book knowledge."

"I like Albert Schweitzer's remark, 'Reverence for life,'" I said. I felt happy —I had come into my own. Alas, not for long.

Einstein looked at me with penetrating eyes. "Yes," he answered slowly, perhaps in consideration for my pen, "when you want to learn about creation, you must have a very humble disposition."

There was a pause, and I asked, "People say that you replaced the geometry of Euclid with that of Riemann."

"Yes, I used Riemann's geometry because he got rid of the idea that the shortest lines joining any two points are always straight lines." Then suddenly his voice took on a mistrusting tone: "What do you know about Riemann?"

I could have bitten off my tongue for having shown off, yet he was a compassionate man and, noticing my difficulty, continued his discussion of geometry.

"Whatever geometrical system man chooses, it is always a construction of the mind and has no connection with reality, for geometry possesses internal order, which seems to be lacking in reality. Reality does not furnish geometry with axioms."

"How then," I asked, "can we use geometry to probe reality when the two are not related?"

"We measure the experience of our thoughts against the experience of our observations. Thus we bring order into the world of reality, and make it comprehensible. But always remember: as far as the laws of mathematics refer to reality, they are not certain; and as far as they are certain, they do not refer to reality."

I asked Einstein to repeat this last sentence, which he did. With a good-natured smile he added, "Now you understand why my friend Max Weber said 'Science is never finished.'"

I heard a woman's voice in the living room, and began worrying that my audience would end before I could trace the genesis of his genius. But my nervousness apparently had no effect on Einstein, who continued puffing contentedly on his pipe.

"What happened after you worked out your formulae?" I asked.

I wrote them on thirty sheets of paper, took them to an editor of *the Journal of Physics* in Berne, went home, got into bed, and was sick for fourteen days."

"Perhaps the editor became sick, too," I ventured, "when he found that it was so monstrously difficult to understand."

"I don't think so; I imagine he scarcely looked at it. Of course, they had published papers of mine before, and it wouldn't have cost them much to print this one also. Besides, I had already grown used to the comforting thought that no one would notice me."

"But this time you threw a stone into the scientific pond," I pointed out.

"No," Einstein smiled. "It was a pebble; it therefore took years before the small ripples at the shore were noticed."

"What will be the practical result of your theory of relativity?" I asked, rather timidly.

Einstein leaned back in his chair, apparently enjoying this question. "Well, you might use the energy of the matter to catapult yourself to the moon, if it becomes too crowded here. Or you could discover how to imitate nature in the laboratory; then cattle could be fed with synthetic material instead of grass. On the other hand, if you are a misanthrope, you can get rid of yourself in a flash and, at the same time, take half the planet with you."

"That's terrible!" I blurted out.

"But you must remember," said Einstein, looking straight at me, "that my formula is neutral. The choice of whether to use it as a blessing or a curse lies with mankind."

"The world will say, There is no end to the name Einstein!'" I said impetuously. He ignored this remark, and I quickly asked, "How long did you work on your formula?"

"For nine years. But it wasn't only one formula. A thousand were written and destroyed; often I wanted to give up. At last, however, after endless labor and innumerable sleepless nights, I found it."

"I have found it, too!" I said, jumping up from my chair. "You have revealed genius to me; it is not created by others, it is self-evolved. If you will permit me, I shall call my radio lectures 'Genius and Tenacity,'* I had rosy visions of the intellectual elite awakening at my call and immediately plunging Hitler's brand of genius into murky disrepute.

Einstein nodded. "It is true that I have stuck to it. My intuition made me work. Many people think," he said assuming the attitude of a lecturer and I again sat down, "that the progress of the human race is based on experiences of an empirical, critical nature, but I say that true knowledge is to be had only through a philosophy of deduction. For it is intuition that improves the world, not just following the trodden path of thought. Intuition makes us look at unrelated facts and then think about them until they can all be brought under one law. To look for related facts means holding onto what one has instead of searching for new facts. Intuition is the father of new knowledge, while empiricism is nothing but an accumulation of old knowledge. Intuition, not intellect, is the 'open sesame' of yourself."

"Could we then say," I suggested, "that empiricism is a horizontal line, while intuition is a vertical line from earth to heaven?"

"What do you mean by heaven?" asked Einstein cautiously, looking at me out of the corner of his eye.

At that moment a lady entered from the living room and invited us to tea. Einstein opened the door for me to pass, but forgot to present me to his wife, so I introduced myself. "I didn't expect such hospitality. After all, I'm only..."

"No, you are welcome," she said smiling.

Frau Einstein looked rather small next to her husband, and her simple grey afternoon dress matched her hair. A plain gold ring was her only adornment.

While we talked she drew close, and I remembered having heard she was nearsighted. She nonetheless seemed preoccupied, even worried.

As we walked through the living room, Frau Einstein invited the sculptor to join us. We entered the library, where books were stacked to the ceiling, and Einstein's secretary was busily arranging the letters and papers scattered over several tables. I noticed there two leather-bound Bibles and a work of Spinoza. The tea table, with sparkling family silver, stood before a large window and formed a pleasant contrast to the somber bookshelves. Frau Einstein served graciously but silently, selecting certain tidbits for her husband.

I learned that the sculptor had been commissioned to sculpt two huge bronze plaques, one for the new Berlin City Hall and the other for the Einstein Tower in Potsdam. Turning to Professor Einstein, I confessed, "This man wants the outside of your head, and I want the inside."

Pointing to the window, Einstein said, "And they want all of my head."

At that moment I became aware of the familiar sound of marching feet. Frau Einstein sighed. Those Nazis. This has been going on for an hour. How glad I am that we are traveling abroad!"

"You are leaving?" I asked, my notebook on my lap, to discreetly jot down the key words.

"Yes," replied Einstein. "I have been invited to lecture in the United States. We leave in a few weeks."

Frau Einstein added, "If only we could stay there."

We all stopped talking at this point and heard distant singing, although the words were indistinguishable. "Who can be trusted these days?" Frau Einstein moaned. "Even former friends seem to avoid my husband."

"It's not surprising," I said, "when a well-known physicist and Nobel prize winner, no less, is heading a movement to disprove his theory." Einstein knew, of course, that I was referring to Professor Lenard.

The sculptor contributed a comment on the recent attempts to attribute Einstein's theories to an Austrian scientist killed in the last war.

Einstein looked into his cup as if for an answer. "I am flattered. Heinrich Heine is said to have stolen his *Lorelei* from another poet who died forgotten." He sighed deeply. "How I abhor these puppets of a dictator! They have their brains in their fists and feet, and the uniform covers their crimes. Mark my words, they will be the men of tomorrow. They will be prepared to hate, and hate is the beginning of war."

I told them how I had once been a part of this hate, how I had marched through the streets of Cologne a few days after the declaration of war in 1914. "We were all civilians: students, apprentices, even schoolboys. We wanted to volunteer for the Army, and we sang with patriotic fury of our hatred for *unseren Erzfeind,* England (our arch-foe). Perhaps German youths never ask questions, but simply follow uniforms and worship heroes."

"Not they alone," Einstein insisted. "They learned from their fathers to bow to any uniform, even a mailman's. Look what Bismarck said: The German lacks civil courage."* After a pause he added, "How true! The German lacks the elementary reaction of conscience."

Frau Einstein almost whispered, "It was this kind of youth that murdered Rathenau."

"I had so many talks with him about Germany and peace," said Einstein. "He was the first victim of Nazi propaganda." His large brown eyes stared at mine. "Remember, he who lives today is responsible for the events of tomorrow."

"How true," I agreed. "Whenever I am invited somewhere I always feel compelled to bring some poems along which deal with the past to shake up the conscience of my listeners and to convince them that we need to create a new man who is more devoted to humanity than to his fatherland."

I then passed around my two poems, "War" and "Verdun."

Frau Einstein was quite moved; handing the poems to her husband, she said, "How gruesome! Verdun, the mother of the French soldiers —I am a mother, too; I know what it is." Selecting some cookies and filling Einstein's empty plate, she sighed with a feeble smile, "Yes, I am your mother also."

After reading the poems, Einstein returned them to me, asking, "Have you published them?"

"I distributed them in the Humboldt Club, the School of Economics and Prisoner of War Association."

The sculptor asked whether Elsa Brandstroem was still the Association's president. "Yes," I answered, "she will be our glorious showpiece as long as the Association exists. I have dedicated my book on my experiences in French captivity to her."

Frau Einstein lifted her cup and said, "I hope you have success in stirring the German conscience with your writings. I've met Elsa Brandstroem once in Red Cross uniform; she was tall, blond and quite pretty. I learned she saved several hundred thousand prisoners of war in Siberia."

I added, "She is called the 'Angel of Prisoners of War!' I am a close friend of hers, and she collects my poetry."

Einstein now raised his cup of tea, "Let's give a toast to women: Madame Curie, who discovers the nature of the physical world, and Elsa Brandstroem, the nature of the beast in man when he imprisons others behind barbed wire."

"And I," sighed Frau Einstein, "discovered the child in man."

"Or the genius," I smiled.

She continued, "Once I had four women to tea. And what did my child do?" she asked, looking over at Einstein without smiling. "He left the table. Later one of my guests excused herself to visit the bathroom and came quickly back, horrified, 'A man is in there hiding behind the curtain!' I calmed her down, saying it was my child who was probably taking a bath."

"Yes," said Einstein, "I can't stand empty chatter."

I turned to Einstein, "How did you first meet Madame Curie?"

"I think it was in 1909. I was invited to Geneva. The University was celebrating its Three Hundred and Fiftieth Anniversary. It had been founded by Calvin."

Frau Einstein interrupted, "And 'my child' told me later that he had thrown the invitation into the wastebasket."

"Because that bulky paper was written in French," said Einstein, "and I thought it was just another piece of mail not worth reading."

Frau Einstein leaned over and whispered to me, "It included an honorary doctorate for my *Albertle.*"

"When I learned of my mistake," continued Einstein, "I of course went and had the pleasure to meet Madame Curie. What a remarkable woman! She possessed great willpower and a lucid mind. The banquet was so sumptuous that I felt Calvin would have burned us at the stake for gluttony."

Frau Einstein again whispered to me, "He did not only feel it; he said it for all to hear!"

There was a pause as Einstein absently stirred his tea with a silver spoon. Then he looked at me, "Why don't you read your poems over the radio. There you have better material to wake the German conscience than to talk about me. And tell your students that humanity is saved by democratic principles such as equal rights for all, the dignity of man, and the right to have private thoughts. Show youth through historic examples, like Goethe, Voltaire, and the prophets of old, that humanity is not to be saved by a dictator."

As he spoke I suddenly remembered the odd couple lurking on the staircase. "Do you know that man and woman who were standing outside?" I asked.

"No," he replied, "but the man accosted me when I went up to my study and was still waiting when I came down."

The maid told me he has been there since morning and wants some manuscript," Frau Einstein said. "I had her tell him he should request an interview."

"If he's still waiting, may I find out what he wants?" I asked.

Frau Einstein replied, "Why should you risk confronting a man of whom you know nothing?"

"Don't go without a gun," the sculptor pleaded. "You don't know who sent him. Friends of mine have disappeared."

"The man looks rather harmless," I said. "Besides, there was a woman with him."

After some debate, Einstein consented to my going out, and I found the couple still there. In the dim light the man moved toward me, probably thinking I was Einstein.

"Sir," I began, "you look to me like an artist. We may have something in common. I am a writer."

"I don't care," he exclaimed. "I want to see that Professor Einstein, and if he doesn't come out in the next half hour, ni blow up the house."

I could hardly believe the man's words, though he sounded as if he meant what he said. I looked at the woman; she was haggard and strained and her face twitched nervously. "Then you would blow up this lady, too," I said, pointing at her.

"I am his wife," she muttered, as if in answer to me.

"What will happen to the dozen other families living in this building?" I continued.

"I shall warn them in time."

At this point I wondered if the man might be drunk. "Oh, you have dynamite?"

"Yes."

"Where did you plant it? In the basement? The janitor who lives there will be pleased to hear that!"

"I have stronger men on my side than the janitor." His eyes flickered strangely, giving him a demented look. He seemed serious about his threats. I asked if there were anything I could do for him.

He pulled out his watch and paced the few yards from the door to the stairs like an enraged animal. "Unless my manuscript is returned to me in half an hour, I will set off the dynamite. Understand? Half an hour!" I turned to his wife. "What's this all about?"

Afraid, hardly daring to speak, she finally told me that six months ago her husband had sent Einstein a drama with a request that he read it and, if he considered it worthwhile, that he write a few lines of recommendation.

The man interrupted, "He never returned it. He probably sold it."

"You know very well that a man like Einstein wouldn't stoop to thievery." Reentering the apartment, I said over my shoulder, "I will see what I can do for you."

The man called after me, "If I don't get it back, there are people who will settle the matter."

I closed the door and took a breath. I thought this man could well have been an erratic disciple of Streicher who had been a writer, too. I recalled how Streicher had founded the newspaper *Der Sturmer,* which was filled with such blood-curdling stories about the Jews that we of the Humboldt Club had sent a delegation to Hitler's office to complain that such hatred would certainly produce catastrophe within Germany and mar relations with other members of the League of Nations. Hitler was reported to have answered, "The Jew is baser, more devilish more decadent than even Streicher's pen could describe him." I had a distinct feeling that it was not by chance that, of all days, the Ides of March, I should climb up the stairs to meet a man who threatened Einstein.

The Einsteins were waiting for me in the music room. I related my conversation with the man, suggesting that he might be a little unbalanced from hunger and worry. Though I suppressed my suspicion of Nazi connections in order not to unduly alarm them, Frau Einstein kept breaking in, "This man wants to destroy my husband!"

"He's a tool," interjected the sculptor. "The testimony of a mentally-irresponsible man could never convict the real conspirators. Paragraph 51 of our law-book will protect him."

Frau Einstein became more and more unnerved. "I hate fame! I hate fame!" she shouted, trembling all over. "Every day there are laundry baskets full of mail, letters in all languages, letters from politicians, aristocrats, scientists, letters filled with insolence, anonymous letters threatening murder."

Einstein grinned, "Yes, one wrote that I had thrown the sun out of its orbit and would be held responsible for the chaos of the world."

Frau Einstein walked in short staccato steps, back and forth, wringing her hands. "Oh, how I hate this fame! How I hate this fame!" Finally she dropped on a sofa, crying, "If we could only leave Germany and hide in the darkest corner of the world!"

Einstein, in the meantime, had taken out his violin and was playing Mozart. "I can stand this no longer," Frau Einstein yelled. "Is there no remedy? Don't you realize that I'm involved, too? Have you forgotten how I was once threatened?"

Einstein kept playing his violin. He probably never heard a word she said, so absorbed was he in Mozart.

She continued, "And to think what I went through when the newspapers falsely reported that you went to Russia! I had a hundred telephone calls."

Half an hour had by now elapsed. I was unable to share Einstein's complacency, and proposed that his wife, the sculptor, and the secretary search for the manuscript and that the maid so inform the man outside. Meanwhile, I suggested that Einstein go to the police. Frau Einstein agreed, fetched me her husband's overcoat, grasped his hand, and pulled him to the rear entrance while the sculptor pushed. It was difficult to persuade him. He seemed utterly lacking in the instinct for self-preservation and was scarcely aware that people could hate him. For Einstein, this threat was as unreal as the ether.

At last we succeeded in getting him through the door. Frau Einstein shouted, "Now, go!" and banged the door shut behind us. I could hear her fastening the bolt. Leaning his large bulk against the wall, Einstein stood motionless, looking down the back stairs in disgust. I became irritated but gently tried to persuade him, "Come, it is better that we go."

He didn't budge. He gazed at me, then at the overcoat which I held ready for him, as if in a daze. I tried to arouse him. "Dr. Einstein, you said regarding physics, 'I don't know any more than a boy\ but with all due respect, let me tell you that regarding people, you know no more than a child either. You have created a mountain of knowledge for us to climb, but you are so helpless. I don't understand you. You are a child, and what a child! I feel I should change the title of my radio lecture from 'Genius and Tenacity' to 'Einstein: The Genius in a Child.'"

He moved away from the wall and I helped him into his overcoat. While we descended, I remarked, "I suppose you've never been down the service stairway before?" Crossing the court, I mused, "Let us hope that the janitor will think we're servants or delivery boys."

He probably didn't hear and set out briskly for the police station, which was about ten blocks away. He said not a word. We were almost totally ignored by the desk sergeants, even though I told our story as vividly as possible. Finally, a tall young man, who proved to be the police captain, entered with some papers, and only then did one of the sergeants approach the banister. I was so irritated that I loudly announced, "I've come to find out whether Professor Einstein is on the black list of any of the fanatic groups."

"Ach," he waved, "how should we know?"

"Don't you remember," I returned, "that Reichsminister Rathenau, also a Jew, was on a blacklist and was killed with hand grenades the very day he was to receive the French and English ambassadors? And now Professor Einstein is going to America, as a German ambassador of science so to speak."

The officer, alerted as to who was there, approached us. "What can we do for you, Herr Professor?" I noticed Einstein was wearing an open-necked shirt, a rather old sweater, and wrinkled pants beneath his unbuttoned overcoat. I thought, "No wonder he didn't make an impression on those sergeants, with their shining buttons and ornate collars."

Einstein tugged at his ear, and with a helpless gesture and an obliging smile, replied quietly, "I don't want you to do anything except make those people go away from my door."

But I wanted them to investigate whether or not the man and his wife had come of their own accord. "We are not interested in vague assumptions," said the captain, whose silver-threaded epaulet and two stars dazzled my eyes.

"Everyone should be interested in defending the Weimar Republic," I insisted.

"Are you insinuating that I'm disloyal?" He lifted his chin.

"Everyone is, who has closed his ears to the sound of marching feet and military songs." I was by now thoroughly provoked and I told them with a loud voice so that the three handsomely uniformed policemen in the room could hear it. "I am concerned about the security of the Weimar Republic. I am a member of the Humboldt Club which was endowed by the Foreign Office to establish peaceful relations among the students of every nation. And the French and British members were incensed not long ago when the Nazi leader Ernst, sporting the brown shirt and swastika, came as a guest of a German student and sat at our table to lunch. The young architect Speer, who remodeled our house, is suspected of being a Nazi. Whom can we trust? Don't you see, gentlemen," I continued, "that the Republic is being undermined? As a member of the Humboldt Club council I proposed that Dr. Einstein address us, but though I had the American, British and French students for support, I was voted down by the German students who were secret Nazi sympathizers.

"You know well that the young German Rudolf Leibus had offered a reward for the murder of Einstein, Wilhelm Foerster and Maximilian Harden, announcing that 'it was a patriotic duty to shoot these leaders of pacifist sentiment.' And how did our courts handle the case? He was fined only 60 reichmarks. Another time I and fellow members of the League for Human Rights, Dr. Einstein among them, petitioned the courts for fair trials of those workers who belonged to the Socialist Party. For months now they have languished in jail because the judges don't dare acquit them due to pressure from the Nazis. When Hitler comes to power they will surely be executed."

The officer was a good listener. He, I discovered by looking at his desktop, owned a monocle, and I couldn't help thinking of the monocled General von Seeckt. He, too, had been appointed by the Weimar Republic but worked to undermine her authority. He had, it was said, even used the funds which Americans sent to Germany for reconstruction to pay for a secret German Air Force training in Russia, thus violating the conditions of the Versailles Treaty.

I felt that I stood in a hotbed of treason. Another policeman rose and came to the banister. His girth was such that I wondered how his handsomely buttoned coat held together. He whispered something in the captain's ear, part of which I caught as "insult to a state official."

This made my flesh crawl. A friend of mine had recently appeared before the police for implying that one of their officers had been bribed by the Nazis. On his way out, he was led into a house where several thugs beat him so badly that he was hospitalized for days. I thought quickly. "I respect the protective uniform," I said. "I, myself, wore one for some five years, and was even awarded the Iron Cross."

As I anticipated, this lustrous piece of my past took precedence over my insignificant civilian present. The captain became less condescending, though he still had no intention of helping us. "Do you expect me to act on the mere assumption of a plot?"

I hastened to answer, "I believe you are right. The mere assumption of a plot is really not enough for you to act on. Don't you think so, Herr Professor?"

Turning around, I found that Einstein had disappeared. With a smile of relief, I said, "Where there is no accuser, there is no case." Clicking my civilian heels, I made a stiff military bow and hurried into the free air.

Einstein stood waiting on the street corner. I was still shaking as I told him of my narrow escape. "And you took me there," he said gruffly.

I resented this. "Professor, those people would have shown you more respect if you'd worn a starched collar and a necktie. They took you for a *Landstreicher* (tramp). And," I quickly added to smooth over my brusque remark, "if you had brought your violin, they would have thought you were a *Wandervogel* (rambling minstrel)."

In the following moment of silence, I realized that I, too, was improperly dressed, for, although I wore my best blue suit, I had forgotten to put on my coat and hat. I didn't even have gloves, in Germany a must for gentlemen, especially in winter.

"Why are you so concerned with what these people think of me?" Einstein snapped, and started home as quickly as before, his hair flying in the wind and his coat hanging loosely on his powerful frame. Walking beside him, I realized that although we had been together for over two hours, he had asked not a single question about me. We exchanged our thoughts about the present danger of Nazism, and then I reminisced about my own war experiences and academic pursuits, adding that presently I was waiting to be called for work with the League of Nations.

Einstein replied, with a sigh of resignation, that the whole League was falling victim to the spirit of compromise. "Can one compromise with rattlesnakes?"

He went on to explain how the Committee on Intellectual Cooperation had in vain proposed international surveillance of the way history was taught in schools.

"That's one of the items we fought for in Geneva," he said, "and we almost had it until Mussolini intervened. The Italians finally agreed to international control only at the university level, believing that students sufficiently injected with Fascist doctrines were immune to pacifism or democracy."

His voice grew scornful. "How I detest this police state. It suppresses political freedom and metes out its justice with two sets of laws. Of course, this perverse reaction called Hitlerism will not last long. I have faith in the dignity of the German intellectuals."

"But, Dr. Einstein," I protested, "did you not experience their nationalistic fervor during the last war? You could not change their minds, as hard as you tried. Didn't you feel isolated, if not betrayed? It was said that in the court of the Kaiser you were called 'a moral leper.'"

At these words, Einstein seemed to bolt ahead of me and I found it difficult to keep up with his long steps. Suddenly he stopped and said, "If you want to help the Germans forget and save the world from a new war, don't count on established religions. We must found a cosmic religion, one of unconditional love, not one which sells itself to those in power."

"Yes," I sighed, "The empress, a deeply religious Lutheran, had a Jewish heart specialist; yet she did everything she could to force the Kaiser to break away from his Jewish advisors Rathenau and Ballin because she was a pan-Germanist."

"Let the dead rest in peace," Einstein replied. "We now must look to the future. What do you think of Spinoza?" he asked, and before I could answer he went on. "For me he is the ideal example of the cosmic man. He worked as an obscure diamond cutter, disdaining fame and a place at the table of the great. He tells us of the importance of understanding our emotions and suggests what causes them. Man will never be free until he is able to direct his emotions to think clearly. Only then can he control his environment and preserve his energy for creative work."

When I ventured to say that Spinoza criticized the Bible too harshly, Einstein glowered. He retorted that I should read him more carefully, that Spinoza rejected metaphysical speculations and explained miracles as natural events, that he was deeply religious but divorced from dogma.

Spinoza's unorthodox theism had shocked me. I asked, "If God reveals himself in nature, why not in man?"

"Have you never been awed by the power of man's rational mind?" Einstein quizzed. "And man's intuition, man's inspiration?"

"Yes, but I am searching for a personal God. I have become a member of the German Occult Society."

"In order to contact God through mediums, I suppose."

"Yes, Professor, I am a mystic. The Old and New Testaments record many of them."

"Ah, yes," Einstein laughed. "King Saul was a mystic, too. He wanted to conjure up the prophet Samuel, so he used the Witch of Endor to satisfy his curiosity. Do you know that Hitler gets his horoscope done by one of the astrologers in your Occult Society? You should look into that. But let us get back to Spinoza. You must study his *Ethics.* I learned from him that there is no time in creation."

"As much as I know," I interrupted, "Spinoza implied that our mind is eternal since everything is God."

"Not the God that speaks through a medium," scoffed Einstein. "Perhaps we are eternal, but I do not speculate about what will happen to me after this life."

"But, Dr. Einstein, you prayed as a child with your mother, as I did, didn't you?"

"We grew up, didn't we? For me God is no longer a father figure. Of course, you are a poet. I'm not interested in what God looks like, but in how the world he created looks. I can read the thoughts of God from nature. The laws of creation interest me, and not whether God is made in the image of man, with a long white beard, and has a son. I am a part of infinity: I see everything in *specie aeternitates.*"

"But don't you personalize God," I asked, "when using such words: 'God is subtle but not malicious?'"

"Dr. Hermanns, I am interested only in living this life according to ethical laws, like Thou shalt not kill,' which often contradict the laws of the State."

Just then, the law of the State became manifest. We had arrived at Einstein's residence when two boys in Nazi uniform stormed out of the building and bumped into us. With a gesture of utter disgust, Einstein ran up the stairs, and I chased after him. He rang the bell while pulling out his keys, and the maid opened the door. From the entrance hall Frau Einstein called, "Albertle, we found the manuscript. The man has taken it and left."

For a moment both Einstein and I were stunned; we had forgotten all about the intruders. As he threw his coat over a chair, his wife said, "The man wouldn't accept our apologies. He said we would hear from him later."

As if it were her cue, the maid dashed from the living room and addressed Einstein, "Professor Einstein, I asked the man, What do you mean we will hear from you later?' He said, 'I have a higher duty to fulfill.' He made a motion as if to pull something from his pocket, but the man's wife stepped forward apologetically. 'My husband,' she said, 'has been in the clinic because of his nerves.' Then she dragged him down the stairs." Einstein ignored the maid, strode into the music room, and soon the house was filled with the sounds of a baroque toccata. Frau Einstein's frightened eyes gradually filled with warmth as I related, in what I hoped was a humorous manner, our experiences at the police station. She didn't smile back, however, nor was she soothed by her husband's musical nonchalance. We could see him through the open door of the living room, playing like a passionate virtuoso, and mumbling now and then to an imaginary accompanist.

The sculptor, meanwhile, was sitting nearby in front of the huge plaque, depressingly silent. His sharp features contrasted strongly with the bushy hair and full-cheeked face of his model. His face and tall stature also set off the soft and motherly features of Frau Einstein. Her gray hair and warm searching eyes were filled with such longing for peace, understanding and love.

"How happy must your husband be," I said to her, "to have you as his wife and mother. At the age of seven I was so lonely: my mother had died, and I soon got a stepmother—a cold Westphalian. A few years later I wrote this poem:

MOTHER

No more the threshold has a smile
and speaks, "Come sup, come rest awhile, my son, don't tarry."
You moved nowhere—but moved—how wrong!
You even took my tears along.

Can you not in a dream appear
to whisper, "Look, my child, I'm here.
Your load I'll carry. *
I feel you not—no touch of love.
Mother, have you no light above?

After listening, she held my hand for awhile, nodded her head and said, "Yes, that must have been providential that I became Albertle's wife after his divorce."

I rose to go. Since she still seemed distressed, I asked, "Surely, you take pride in your husband's accomplishments, don't you, Frau Einstein?" She looked at me as if I spoke in a foreign tongue. "Frau Einstein," I murmured, "What do you know about relativity?"

"Oh," she said, waving her hand as if to dismiss the question, Tm always asked that. I know nothing about it. I don't need to, but I know my husband." Her clear azure eyes looked at me with concern.

"You mentioned," I said impulsively, "that you had once been endangered. May I ask how?"

"I still tremble when I think of it," she began, wringing her hands. "Some five years ago, a woman came to speak to my husband. When I asked the reason, she refused to answer. Sensing danger, I called the police, but in a flash she was on me with a large hat pin. We were still grappling when the officers came and took her away."

Reliving this experience, she became distraught, and leaned forward to touch my arm. "Can't you persuade him to leave Germany for good?" she pleaded. I was amazed at the despair behind her appeal. As if reading my thoughts, she said, "He runs toward danger with his eyes closed. I almost went out of my mind after Rathenau's murder when he rode in an open car through the streets of Berlin to protest war. He's such a child! What can I do with him?"

The sculptor laughed quietly, saying, "It's not easy to tell your husband what he should do." Then he looked at me. "When a friend tried to persuade him to leave Germany, he answered, 1 can't see that I'm in danger. I still feel like a man sleeping in a comfortable bed. Of course, I must admit that now and then I'm bitten by a bedbug.'"

Frau Einstein stepped toward me and, with her hands clasped together at her breast, said, "I implore you, don't bring my husband into your radio broadcast. They may put you on the list, too."

"We must open the eyes of the German people," I replied.

Frau Einstein shrugged her shoulders in defeat and went to tell her husband I was leaving. The sculptor added, "Of course, she's nervous for a good reason. On one occasion, the Einsteins arrived home late from a play and noticed two suspicious-looking men waiting in front. So they rode around for half an hour, and upon returning found the men had gone. When the janitor was questioned, he said they claimed to be students of Professor's, but, of course, no one believed it." Pointing to the plaque, he said with a sardonic smile, "I wonder whether this bronze will be put to another use. The students at Berlin University tell a story of a professor who ordered a Christmas play-box from Nurnberg for his daughter. On the outside was a beautiful picture of a doll's carriage, but when the parts were assembled, it turned out to be a machine gun!"

Einstein, entering, overheard the last words, and said to the sculptor: "Yes, I must wake up—live more in this world than in my laboratory." He accompanied me to the door. I had taken a step outside when an inner voice caused me to ask, "Herr Professor, will you please write your name and the date on a piece of paper?"

We returned to the library, and Einstein pointed out a shelf filled with Heine. "Thor will leap to life and smash the Gothic cathedrals with his giant hammer.' These words, which Heine wrote a hundred years ago, will soon be fulfilled. Hitler will smash the Gothic cathedrals." He took a deep breath. "It's sad. What can we do?" Searching for a piece of paper among books and files he mumbled, "You, I must say, came the right day for me. I now see what is afoot."

At last he found a theatre invitation and wrote on the back: "In memory of the incident. March 4, 1930. Albert Einstein." I was elated that this eminent man, normally so unconcerned about events in his personal life, should have documented this sinister occurrence. "These five words, Dr. Einstein, shall be a testimony to the world of what a noble mind has to suffer and how you were treated by the police."

Einstein accompanied me out of the apartment into the corridor, where we heard from afar loud, vigorous singing. Although the marchers were blocks away, the sound wafted over the gardens in the rear of these elegant apartment houses in Berlin's Bavarian section and drifted through the open window on the fourth floor. We stood listening and watching for what seemed an eternity. It was as if we were latecomers in a theatre and didn't move for fear of disturbing the drama. Who were the actors? Germans. Who was the audience? Germans. And yet from the stage came a devastating hate specifically directed at us.

"Dr. Einstein," I said, suddenly gripped with fear, "that man on the stairs may have had a pistol in his pocket; maybe he was to receive a reward for killing you. Do you hear what they sing now? It's what we sang marching to France in 1914:

Red of the morning, red of the morning,
your light brings us early death.
Yesterday mounted on proud steed,
today a bullet through the breast.

Einstein whispered, "What a betrayal of man's dignity. He uses the highest gift, his mind, only ten percent, and his emotions and instincts ninety percent."

"Goebbels is responsible for this!" I continued. "He is the one who has aroused these youth. A baker's apprentice, a boy of sixteen, was murdered in a street brawl while defending his faith as a Catholic. Since he had lived on my street and delivered fresh bread to my landlady every day, I went to pay condolences to his mother. I found his body in the coffin, with four young boys dressed in brown shirts standing guard. Thus Goebbels, the Gauleiter of Berlin, who had first encouraged street brawls, also sent his condolences! Who knows whether, after an insane man has been turned loose on you, Dr. Einstein, you, too, will be so honored at your funeral."

Einstein seemed to have shrunk as though he wished to hide himself. His broad shoulders were drawn together, and his head bent forward as he gazed up the stairs leading to the attic. "There is my refuge," he stated simply. "There I'm happy."

"I believe you, Dr. Einstein. Hate can never destroy your laboratory and formulae." Suddenly I clutched his arm. "But now I know that the League of Nations is dead. They didn't stop Musso-lini, and they will be all the more powerless against Hitler."

Einstein nodded. "It was a dying child when it was born. It had no instrument of power to reinforce its laws." Seeing my chagrin, he advised me gently, "Forget your dream of working for the League. Instead, set your mind on reality. Rally your generations, show them the hideous effects of Nazi propaganda. The older gen-eration betrayed youth in the World War, and now, once again, they are leading the sheep to the slaughter. You've seen the Battle of Verdun. You know how it is."

I gave him my hand and promised to do my best. "And maybe we can address a letter to President Hindenburg."

"He wouldn't be the man," said Einstein. "He knows how I de-spise the military. To kill a man on a battlefield when you don't even know him is just as much murder as if you were to shoot a citizen on the streets of Berlin. One crime is rewarded with medals and glory, the other is punished."

There was a creaking sound behind us as the door opened. Frau Einstein said quietly, "Albertle, isn't it too cold outside?"

He seemed not to hear. The door closed. His eyes were drawn again toward the window where the echoes of marching feet seemed louder, and for a moment his face became contorted. He turned his head quickly toward his apartment, stepped backward and stopped. "No wonder German youth behave this way," he said. "I'll never forget how a teacher of mine took a long rusty nail from his pocket, held it up, and said, 'With such a nail the Jews crucified Christ.' Although I was very young at the time, I was already feel-ing the tragedy of being a Jew. That was in a Catholic school; how much worse the anti-Semitism must be in other Prussian schools, one can only imagine."

He sighed and wiped his hand over his forehead. Did he want to be rid of sweat, or memories? I gazed through the window to the street below, where hundreds of boots pounded the pavement like hammer strokes. The sound of each booming step raced up my spine. I commented, "And to think that the parents of these boys are churchgoers and that the youth have had religious instruction in their schools! I went as member of the Humboldt Club to the Reichstag a few months ago to find out more about the position of the Democrats, the Socialists, the Catholic Center Party and others who belong to the majority coalition. I tried to get a speaker for our club, but they are all afraid of later revenge by Hitler."

Einstein stared at me with his large brown eyes, "The Center Party? When you learn the history of the Catholic Church, you wouldn't trust the Center Party. Hasn't Hitler promised to smash the Bolsheviks in Russia? The Church will bless its Catholic soldiers to march alongside the Nazis."

"But Dr. Einstein, a new bloodbath?"

Einstein smiled mockingly, "Didn't the Church bless the soldiers who went with God, Kaiser, and Fatherland?"

Einstein drew himself up as if he were a Biblical prophet, "I predict that the Vatican will support Hitler if he comes to power. The Church since Constantine has always favored the authoritarian State, as long as the State allows the Church to baptize and instruct the masses."

I grabbed his hand. "Dr. Einstein, I fear the worst."

Einstein nodded his head, "Heine also feared the worst when he wrote that Christendom was only able to soften the Germanic war spirit, but could not destroy it. As soon as the tamed talisman, the cross, crumbles, the furor of the old fighters, that mindless berserk fury, rattles again. 'Thor will leap to life with his giant hammer and smash the Gothic cathedral.'"

"Heine is right," I said, "Goebbels' newspapers are reflecting this Teutonic berserk fury."

I felt Einstein's eyes piercing my own. "Dr. Hermanns, don't go to the radio to speak of me —I have a feeling you endanger yourself."

"But I made a pledge on the battlefield of Verdun that if God would save me there, I would serve him as long as I live. I now believe in a personal God, who doesn't limit himself to speaking through cosmic laws, since I was saved four times on the battlefield after this vow. I am a metaphysician."

Einstein smiled, "And I a tamed metaphysician. I can only say what I feel. A new world war will come under Hitler. I can't say when, but I feel it will come. This phrase-mongerer Hitler feeds the Germans with the Teutonic beliefs of they being the superior Nordic race. And, of course, the Germans like nothing more than to look up at authority; then they can express their true instincts: 'March, cry for Battle, and kill. Germany over all!' Therefore, the Weimar Republic, built on democratic principles, will not last. The cosmic laws will take care of Germany. Later generations will reap what Hitler has sown." He looked at me with his large brown eyes, "Maybe my wife is right, we should not come back from our trip to America. No, I have to come back. I still believe that there are good Germans."

What a man, I thought, as trusting as a child. "Good Germans?" I asked. "Didn't you say to me, that in times of crisis the Germans make decisions without their conscience?"

As if a higher power supported my words, just at that moment the chant rose through the open window, "When Jewish blood spurts from our knives, then it will be twice as good!"

"These are the good Germans of tomorrow," I said, pointing below.

Einstein stood there for a moment as if shaken out of a dream. He threw his hands up to cover his ears, turning his head toward the window. Then he looked up to his attic study, and I wondered if he would rush up there to bury himself in the familiar world of his books and formulas. But he did not move. He took his hands from his head and looked at me, pausing as if to find the right words, then he whispered, "I feel that it will not go well for the Jews." Putting his hand on my shoulder, his eyes piercing mine, "Can we still do something?"

He paced back and forth from the window to the door, mumbling, "What can we do, what can we do to save the Germans?"

He raised himself before me, looking so sad that I thought he may break out in tears any moment. "Herr Hermanns, it was not by chance that you came to me today that we both had to witness this. The hour is late, but we must try. We must change the heart of man. If something happens to me, promise that you will continue."

I wanted to give him my hand, but he had already turned his back, pulled out his key, opened the door and closed it behind him.

I stood there stunned. It was as if I felt the vibration of that man whom I had met here on this spot some hours before. I stood there as the words pounded within my head: "We must change the heart of man. And if something should happen to me, promise that you will continue."

That Einstein, who never seemed to care for his own safety, should tell me this showed that he himself was not immune to his experience with the distraught man, nor of the hate sung by the marching youth.

Yet, in spite of all, I felt rich. This foreboding day was to give me one of the greatest treasures in my life—a card with five words penned on it: "In memory of the incident. March 4, 1930: Albert Einstein."

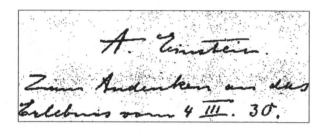

Note signed by Einstein in memory of the police station experience, March 4, 1930.

The Second Conversation

Introduction

On the 24th of January of 1934, I crossed the Rhine into France with the Gestapo at my heels, a refugee like Einstein. After a three year search in over half of the world for a new Fatherland, I received from my sister Hilda, who had married an American, an immigration visa, which enabled me to become an American citizen. And more —it enabled me to again come into personal contact with Einstein.

I had reasons. Important things had happened which justified, or so I thought, my attempt to mobilize him once more against Hitler, who was relentlessly drawing the whole world into war to fulfill the proclamation of a group of German "Christians" in April 1937: "Hitler's word is God's law, the decrees and laws which represent and possess divine authority." Hans Kerrl, Reichsminister for Church Affairs declared, "Adolf Hitler is the true Holy Ghost."(1)

As if to obey C. F. Jung's principle of synchronicity, or meaningful coincidences, one of my first encounters on American soil in 1937 was in New York, with a German student who invited me to visit the German Bunde. In the short time I was there, I witnessed something of the spirit that attracted its members. While they sat in their club chairs drinking beer, colored pictures of Hitler were passed around showing him in the silver garments of the Knights of the Grail. It was this Hitler of whom the renowned sexologist Magnus Hirschfeld has stated that it was his natural perversional tendencies of masochistic degradation of women in his sex play, as well as embarrassing homosexual contacts in his youth, that forced him into the successful compensating acts of the outer man, especially evident in his book *Mein Kampf*, which mesmerized the Germans who circulated the words, "We believe in Holy

Germany, Holy Germany is Hitler! We believe in Holy Hitler."(2)

What Magnus Hirschfeld had told us about Hitler in our confidential circle of supporters of his Institute for Sexual Enlightenment had prompted me to visit in 1932 the famous painter Max Liebermann, who, having painted von Hindenburg, would have been an excellent messenger of what Hirschfeld had told us. However, it became clear to me during my visit that Liebermann felt himself watched by the Gestapo, who indeed would have already sent him to a concentration camp were it not for the patronage of the still living President. Nonetheless, his paintings in the Berlin Museum were thrown out.

I tried also to influence the dramatist Gerhard Hauptmann when I was among the guests at his seventieth birthday celebration in the Berlin State Theatre, a performance of one of his famous plays, Gabriel Schilling's Flight. I went to his box during the intermission, but I had no chance to speak my mind. The conversation stayed as immaculately formal as our attire—his wife wore an orange silk dress with diamond ornaments—and although Hauptmann was President of our Writers' Guild, which collected my poems, he insisted on speaking about the two main actors in the play, Werner Kraus and Elizabeth Bergner.

My search for an ally in my fight against Hitler led me in 1932 to the second wife of the Kaiser, but my appeal to the Empress found a catastrophic ending at an evening soiree on February 28 at the Crown Prince Palace on Unter den Linden. The Empress revealed her intention to place her husband, the former Kaiser Wilhem II then living with her in Doom, Holland, back on the throne, and she asked us to vote for Hitler in the elections to be held in two weeks. Her guests were mostly high-tided aristocrats, generals, and specially chosen people like me, a poet and young diplomat, who could perhaps influence our professional circles, and she moved from group to group exhorting them, "When Hitler has cleaned the Augias stable of the Communists, then we come back!" I, having come to enlist her against Hitler, ran after her just as she was entering another salon to spread the word to more guests, almost tripping over her bejeweled crimson velvet train in my haste. My anguished words, 'Your majesty, Hitler is an evil man!" produced a shocked look on her face, changing into a smile, with her remark, "My dear Dr. Hermanns, if there were no evil persons in this world, then we would all have angel wings and fly somewhere else."

When, on the very next night, the Reichstag, stronghold of German democracy, went up in flames, I knew with certainty my days in Germany were numbered. A short time later thousands of books were publically burned in a tremendous bonfire in front of the University of Berlin, just opposite the Crown Prince Palace. What meaningful events! My manuscript had been burned, too.

In America, with these experiences in mind, my thoughts turned toward saving what would be left after the war already in progress by getting Einstein, as it were, on the bandwagon of founding an international police force. In 1943 students from Princeton, New Jersey, who attended Harvard where I was lecturing at the time, told me that Einstein was sick and could not receive anyone. When I personally wrote him a letter reminding him of our first meeting in Berlin in 1930, and sharing one of my experiences with faith healing, he wrote back on March 23, 1943:

Dear Mr. Hermanns:

The scene in my house in Berlin, which you have referred to, was not a putsch, but a visit of a mentally ill author, who was angry at me, because for a long time, I could not find his manuscript, which he had sent to me.

To your other subject, I can only remark, Belief makes holy, but only to them who believe, therefore not me, but only you.

Friendly greetings,
Yours,
A. Einstein.

March 23,1943.

Herrn William Hermanns
8 Bryant Str.
Cambridge, Mass.

Sehr geehrter Herr Hermanns:

Die Szene in meinem Haus in Berlin, die Sie erwähnt haben, war kein Putsch, sondern der Besuch eines geisteskranken Autors, der gegen mich aufgebracht war, weil ich damals sein Manuscript lange nicht finden konnte, das er an mich gesandt hatte.

Zu Ihrem andern Gegenstande kann ich nur bemerken: Glaube macht selig, aber nur die, die glauben, also nicht mich, sondern nur Sie.

Freundlich grüsst Sie
Ihr
A. Einstein.

Letter from Einstein, March 23, 1943.

Typical Einstein! He would never go further than to say, when his life was endangered, something like this: "Yes, it is true, I was sometimes bitten by bed bugs, too." Or, as his wife told me, "Albertle is a child and will die a child."

In the summer of 1943 synchronicity arranged the events which would put me at Einstein's doorsteps in Princeton when at the annual meeting of the Cambridge Christian Science Study Group a Harvard professor intimated that Einstein shared the Christian Scientist concept of a universe of spirit, not matter. Wishing to help the organization that had so many times assisted me in my distress, I set out to get a statement from Einstein to that effect. The lady who volunteered to drive me to Princeton commented, "You know, matter isn't real; Mary Baker Eddy has proven it." Her offer was the principle of synchronicity at work again —an outside crutch to help me to bring my inside pleas, accumulated over ten years, to Einstein, my "spiritual father."

References:

1. Langer, Walter C, *The Mind of Adolf Hitler,* Basic Books, New York, 1972, p. 36.
2. *Ibid,* page 56

Second Conversation With Einstein

August 1943

The Institute for Advanced Studies is in a large meadow, isolated from the town and University. A secretary led the way to the rear terrace, where the peace and beauty of the landscape soothed my mounting tension. The mist lifted its blue veil, revealing the belfries of the University in the distance. A shirtless young boy mowed the grass. Before me the sun pushed its bright fingers across the green, velvety lawn toward the woods beyond, filling them with a golden glow.

A grey-haired man dressed in white appeared and introduced himself as the Institute's Director. He left after briefly interrogating me and soon returned with another man. For a second, I thought he was one of the gardeners. He was rather stout, with baggy trousers and a green shirt, open at the neck; then I recognized Einstein, his face framed by long hair, its silver fringe touching his shoulders. Extending his hand, he said, "Yes, yes, we know each other," whereupon the Director withdrew. "Let us go into my room. We shall speak German together."

I heartily agreed. It was the language in which he had written his theory of relativity, and I, many poems. As we walked through a long corridor, I wondered how best to fulfill my mission, for I had much to discuss: the fate of immigrants, an international police force, communism, and the philosophical theory that there is no matter. I felt a slight twinge of anxiety, because I felt I was in the presence of my spiritual teacher.

The furniture in his office was simple and functional, with only a few pictures and a large table. The wall facing the garden was all windows. We sat down, and Einstein waited for me to speak. I scarcely felt the nearly thirteen years and three thousand miles which had intervened between this conversation and the one in Berlin. His face might have become more wrinkled, his hair had turned white. Though he was sixty-four, his youthful spontaneity was the same.

I took out some paper. "Should I have to quote something to my students, I wish to be correct," I explained.

"You mean misquote." Einstein smiled mischievously and paused. "How long have you been in the States?"

"Almost six years."

"I gather from your notes that you were in danger, too." He referred to the two messages that I had recently sent him through students, preparing for this conversation.

"But I'm not as important," I said. "The Nazis haven't put a price of 30,000 marks on my head."

Einstein grinned and touched his head with his right hand. "I haven't felt it, and it hasn't interfered with my thinking."

Outside the window, a slender grey-haired woman looked toward us as she passed. She resembled Mrs. Einstein, so I asked how his wife was.

Einstein looked at me for a moment. "My wife died seven years ago."

I could have pulled out my tongue. But then I thought that Einstein, with his all-embracing love, probably did not mourn long. "Yes, I forgot," I stammered. "I read about it in your biography."

"My biography?"

"Certainly," I replied, and named the author.

"How silly!" he exclaimed. "How can one who knows nothing about me write my biography?"

"Aren't you interested in reading what people write about you?"

"Not at all." He shook his head. "I shall not read it; and if I had my way, there would be no books written about me. I have never found myself that interesting."

After a long pause, I said, "I also visited Max Liebermann on his eighty-fourth birthday. I wanted to have him approach President Hindenburg in the name of the League for Human Rights to mobilize his conscience against Hitler."

Einstein smiled sadly, knowing my visit was in vain.

I then spoke of the meaningful coincidence that happened on

my visit in the form of a birthday letter from Gerhart Hauptmann: "This letter gave Hauptmann away. He wrote that his long time friend Liebermann, as well as he, should withdraw from public life and leave it to the younger generation. Liebermann tore the letter in pieces, crying, What a hypocrite! Hauptmann has embraced the Nazi doctrine lest Goebbels bar his plays from the German stage.' I picked the pieces up, for Hauptmann was, after all, the president of the German Writers' Guild of which I was a member, trying to save them for me as a testimony to this day. Liebermann grabbed the scraps from my hand and threw them into the fire."

"What a shame," said Einstein, "that Hauptmann stayed on while all the other great writers like Remarque, Thomas Mann and Werfel left. Liebermann's will be the greater glory, even though the Nazis threw his paintings out of the museums and art galleries as 'non-Aryan' art. Fame and honor do not always go hand in hand."

I told Einstein of my experiences with the Empress Hermina in Berlin. When I came to the place where I said, "Your Majesty, Hitler is an evil man," Einstein rose from his desk; he sat down again. "How naive can one be?" Throwing up his hands he said, "And you are still alive... What did you wear your uniform with the Iron Cross?"

"No," I replied, "my evening coat, and I did not wear the ribbon of the Iron Cross, because as you know I had become a pacifist. There were others in civilian suit, and I believe I saw Fritz Thyssen."

Einstein interrupted, "The industrial magnate; of course, he would be there in such a gathering sponsoring Hitler, for that spells war; and Thyssen, Krupp and Stinnes want to sell their wares." Einstein chuckled, "And they wanted you to be part of this conspiracy?"

"Yes," I said, "this gathering had a beautiful name, 'Empress Her-mine's Charity Bazaar.' And when I came in she immediately showed me my poems that I had sent to her among the many gifts spread out on the table."

Einstein broke in, "Your war poems, too, on Verdun?"

"Yes. She knew that I was a pacifist, but she was a broad-minded woman, and I learned that very evening from a count and former courtier of the Kaiser that the House of Hohenzollern needed to make an alliance with Hitler. Hitler knew too much of their private lives and of a certain Osthilfe scandal, regarding acquisitions of some of their big estates in East Prussia, and would blackmail them by saying, 'Either you will do my bidding and vote for me now, or when I come to power I will take away all your lands.' Even the Crown Prince, who had close Jewish friends, made it known that he would vote for Hitler."

Einstein nodded, "The simple-minded win the day. You gave her soul some food, just as Hitler nourished her finances."

I responded, "Dr. Einstein, you agree that she is a noble woman. We both had a common friend, Professor Grutzmacher, the theologian. He told me that she cried a tear when he read to her the poem about Verdun. You know, war struck the woman very hard, ever since her first husband was killed in the First World War."

Einstein paused for a moment and then said, "The God you prayed to at Verdun must have helped you out of Germany. That assembly of Prussians and the Empress needed Hitler to renew the glory of Frederick the Great, and they were willing to put their sons in uniform to see it happen. Considering your remark about Hitler, you were very lucky to get out unscathed."

"Prince Hohenlohc, who was also present at the Charity Bazaar, told me later in Paris that he believed only five of us escaped the Gestapo." I rose from my chair. "You warned me in 1930 that I shouldn't broadcast about you and Hitler or my life would be endangered. You were right. A few years later, I went to the Berlin Broadcasting Company with a radio script on Pergolesi, whose two-hundred-year-old musical creations were enjoying a renewed surge of popularity. To my surprise the station's manager, Kutscher, offered me a high position, because Professor Dessoir had just been dismissed due to his Jewish blood. When I told him that I had a non-Aryan grandmother, too, he looked at me, stunned, and said, 'We wanted to honor you as a war hero of Verdun, but now I remember, your broadcast "Genius and Tenacity" made subtle contrasts between Einstein and the Fuehrer. It was daring to insist that a genius is made from within and not through the acclaim of the masses, but to come here is foolish, Dr. Hermanns, for non-Aryans are not even allowed to enter the building. Didn't you see the uniforms of the door guards? Everyone here is sporting a swastika, even I, though I conceal it under my lapel. I cannot even let you out of here alone: they may ask you for identification at the door. Yesterday they turned a man in to SS headquarters, saying that he had been a communist spy.' Director Kutscher then led me out through a back door where cars were parked."

Einstein nodded, and I sat down. He played with his pipe in silence for a while, and I continued. There was another time that my life was endangered because of my poetry. Remember how you told me to circulate my poem War among the youth? Well, I took it to the Humboldt Club, and what happened? One of my fellow Humboldt members gave the poem to his brother in the SS. There the poem won acclaim among many high Nazi officials. They said that never had war been so glorified! When I told this member, who was destined to be a diplomat himself, that I wrote the poem as a solemn protest against war, he said simply, The German has no subtle mind. For him something is either white or black."

After some more puffing on his pipe, Einstein said, "Yes, the God who saved you at Verdun is still working. Oh, Dr. Hermanns, I'm an old Jew, I believe in the God of law and justice. The Germans cannot win this war," he shrugged and sighed, "or our whole creation doesn't make sense." He chuckled, "I was Hitler's public enemy number one."

I clapped my hands, "Bravo, and I was number one in the pocket edition."

Einstein laughed heartily, and said, "Laws cannot be brought under statistical formulations, especially dealing with race, and certainly not He who created the laws."

"Do you know," I told him, "before leaving Germany I walked by your house at Caputh, after it was raided by storm troopers, and didn't dare even mention your name. The Nazi newspapers had reported that they had discovered in your isolated lakeside cottage 'machine guns, bombs, and ammunition for the Communists to use when they took over.'"

Einstein laughed. The papers failed to mention that the most terrible weapon I had there was an old bread knife. I never dreamed that my modest home would be honored by such well-groomed gentlemen as the Gestapo, when before it had only been visited by thieves in the night. Of course, it was a pretext to seize my bank accounts. Let's be glad we're in America."

"Your life was continually threatened even after you left Germany," I reminded him.

"My life threatened?" He was amazed. "I don't know anything about that."

"But it was, several times. If the Belgian government hadn't had you under guard or if Scotland Yard hadn't checked every step you made, you wouldn't be alive today. Even as you boarded the ship to America, carrying your violin as if nothing else in the world mattered, Scotland Yard was with you."

We both laughed. When I reminded him of the half-starved writer who appeared at his apartment the day I had been invited for tea, he said, "I don't know whether or not I was in danger. But I didn't wish to place anyone else in danger. I hate cruelty of any sort, especially that perpetrated by the State, that 'brutal mechanism with no head but a million feet.7"

"But how can you expect justice, if you gloss over your own terrible experiences?"

He was amused. "You don't expect me to believe in justice in a State where every official, down to the last postman, wears a uniform, do you? The older I become," he mumbled, "the more I wish to be left alone."

I nodded my head. "By the end of 1933 it was obvious that my days in Germany were numbered. But I needed a passport, and since I knew a few influential Aryans in the Rathenau Society, I attended the next meeting to present my predicament. The old servant in the Rathenau mansion who took my hat and coat was dressed in the livery he had worn, or so he said, while his master was still alive. As I passed a priceless Greek statue, I wondered how long it would be before Goering confiscated it and the other treasures on the estate. I was shown into the salon, whose elegant gold and green decor contrasted strangely with the drab dress of the gentlemen who comprised this solemn gathering. Everyone seemed preoccupied and time-conscious, but soon we proceeded to the library and sat at a huge oval table. We discussed the past distribution of Rathenau's prophetic book In days to come to university students, in which the millionaire industrialist and political advisor to the Kaiser advocated consumer-producer guilds in a decentralized, democratic social order."

"That was his undoing," said Einstein. "He was a man of conscience."

"As we were sitting there the door was suddenly thrown open: at last, Mrs. Andrea, the murdered Rathenau's only sister and heir. Dressed in black, thin and erect, she stepped into the room holding a letter in her hand. Signed by Buergermeister of Freienwalde, it read, as I remember, something like this:

To Mrs. Andrea, born Rathenau:
The Board of Trustees of the Rathenau Foundation, City of Freienwalde, have unanimously voted to exclude you as a Board member. This coming Saturday night the people will march in a torchlight procession through the Castle Park to honor the Fuehrer. Your husband Mr. Andrea is invited to attend.

"Nothing moved but the hands of an old clock and the tree skeletons outside in the wind. Someone broke the silence: 'I believe this is the end of our meeting and the end of the Rathenau Society.' Watching Mrs. Andrea leave, I felt the full force of the insult: she, the trustee of a fortune in cultural treasures, dismissed from the foundation which she had created, but her non-Jewish husband invited to attend a Nazi rally on the grounds of the confiscated estate. I lingered a while in the magnificent library. The old servant, helping me with my coat, said tearfully, This is the end for me, too.' Not far from the house, I stopped to look at the place where the assassins had forced Rathenau's car to the curb in 1922 and had blasted him with a hand grenade and bullets. I stood as if nailed to the spot. I felt then that blood and infamy was in store for the German nation. Hitler was bringing the people into the clutches of the powers of evil."

Einstein nodded. "This illustrates the depravity of the Nazi mind. *Pecunia non olet.* "After a pause, he continued with a stern expression, "The money of the Rathenaus didn't smell, but that committee in Freienwalde did. What a pity that Rathenau accepted the vulnerable position of a leader, instead of guiding the Germans from a desk with his brilliant mind. He deprived the aristocrats of the privilege of controlling German politics."

Einstein now looked contemplative and resigned. "A Jew never learns," he continued. "His idealism leads him again and again to pull chestnuts out of other people's fires. So often in history the Jews have been instigators of justice and reform whether in Spain, Germany or Russia. But no sooner have they done their job than their 'friends,' often blessed by the Church, spit in their faces."

I continued, "Through Cabinet Counselor Woldt, a friend of Rathenau, I obtained a passport and visa to leave Germany on the pretext of observing the International Labor Office in Geneva, where I had qualified for a position under Albert Thomas but had no more hope of assuming since Hitler pulled Germany out of the League of Nations some months earlier. At the border station of Kiel, however, I was interrogated by a guard and forced to accompany him to the baggage car to inspect my trunk with my forbidden manuscripts on Verdun and my conversation with you, Dr. Einstein, in 1930. On the way we suddenly heard the screaming of women, whose husbands were removed from the train to face some charges, and my guard jumped down to help push the women back into the cars. At that, the whistle blew. As we passed by the *tricolore* at the middle of the Rhine Bridge and as the tall spires of the Cathedral of Strasbourg loomed into sight, I sank to my knees: I was saved from a concentration camp and death. My vow at Verdun was still potent."

"Yes," nodded Einstein, "you were most fortunate. How despicable people can be when a dictator is their savior!"

I went on with my odyssey and told him that I had first emigrated to the Union of South Africa, thinking that 3000 miles would be a barrier between me and my persecutors. "But I was mistaken," I said. "General Smuts even begged me to stay, and said that as long as he lived there was no danger for me or any other German immigrant. I asked him how much longer he expected to live —he was already seventy —and whether he knew that the Boer youth were marching and drilling in uniforms with swastikas."

"I can understand why you left South Africa," replied Einstein. "Hitler started his party with only a few men."

He looked at me through the large window, and after a long pause, he said, "I will tell you why the Weimar Republic has failed. It was not because of the Constitution, but because the people had no faith in a government led by men who weren't barons or counts or generals. The Americans have chosen better: their democratic institutions are headed by a President who has great power during his term of office, but is at all times responsible to the people. The German people, however, expect no political responsibility from their rulers. Their Untertanengeist (servile mind) asks no questions, because a German is trained to obey the State rather than his conscience. When I was sixteen and went to school in Switzerland, I tore up my German passport and became a Swiss citizen.

"The German mind is a product of German history; and one cannot change the past. For centuries this mind has been formed by schoolmasters who ran schools as if they were military camps. I can still see my teachers, who bellowed at us boys in the school yard when we didn't march in equal step into class. This military nature enhances their personal stature at the expense of the brain.

"I don't believe it's possible to make a German think along democratic lines," Einstein continued. "If he does, the thoughts come from without, never from within." He chuckled to himself and leaned his armchair back on two legs. The German mind is unpredictable and enigmatic. I am a foul-smelling flower or a thistle, and yet they keep tucking me in their buttonholes. What a tragedy when I think of Planck, Heidegger, Jung and the rest of German thinkers. It reminds me of Nietzsche, when he says, 'Culture is, above all, unity of artistic style in all the manifestations of a people's life. But to have learned and to know a great deal is neither a means to culture nor a sign of it, and when necessary is perfectly compatible with the opposite of culture, barbarism.'"

"But why did you later become a German citizen again?" I asked.

"This came along with my appointment as Director of the Kaiser Wilhelm Institute. I had two passports for a while. It is strange how life is. Although the Kaiser confirmed my appointment to his scientific institute, I don't believe I was persona grata."

"How could you be! You protested against the invasion of Belgium in 1914, and some of his courtiers called you a 'moral leper.'"

"It is also interesting," Einstein continued, "that the two advisors the Kaiser most trusted were Jews: Rathenau and Ballin. I met the Kaiser once. He made the impression of a good man who ratled his sword to please others."

"The Empress wanted me to bring you along to the next gathering."

Einstein laughed, "She wanted me to vote for Hitler, too?"

"I imagine so," I said. "She mentioned the round-table conferences of Frederick the Great, who was a great supporter of scientists and artists."

"Yes, I know. He, too, had a Jew at his table, the grandfather of Mendelsohn, the philosopher."

"People in Harvard have asked me," I said, "why you didn't devote your time to investigating the German mind. They felt such a study would aid those who must deal with Germany after the war."

Einstein threw up his hands and said, "Leave me out of this!"

After a moment of silence, I asked whether he had read the report I had sent him on the fate of the German immigrant in America.

"Papers in my study are piled this high—", he raised his hand about a yard above the desk, "and I know that new Americans don't always have an easy life here. Yes, I realize that immigrants have trouble adapting themselves to the language and customs of a new country." He smiled. "But our countrymen seem to be quite touchy: first it is too hot for them, then too cold, and they talk as if all they ever had in Germany was Riviera weather."

"Dr. Einstein, you don't seem to get my point. I remember reading of an immigrant who wanted to photograph you. He was shown out by your housekeeper, who said no one could see you without an appointment. He was persistent, and during the dispute you appeared at the door and the man cried in despair, 'Professor, you are a refugee as. I am. And I must make a living. I was told that, as a refugee, I might be able to do what Americans cannot do: take your picture.' And you allowed him to photograph you. This is my point: an immigrant must do more than an American if he wants to succeed. And if he has a foreign accent and is not good-looking, he must do doubly well. Of course, genius gets its due, but I'm just little Willi."

A wave of his hand ended this topic. "How are you getting along at Harvard?"

I replied, "I had made the same mistake of so many others in thinking that it is better to be the last at Harvard than the first somewhere else. Despite numerous recommendations, Professor Sorokin, the head of the Department of Sociology, told me, 'If we wanted a new man, we'd choose a refugee who already has a name and there are several of them who linger around Harvard Square with empty stomachs, looking for work.' I am doing some research and lecturing there, but the pay is poor."

Hoping to bring the conversation back to my topic, I added, "I feel that I am qualified to serve as a mouthpiece for the immigrant. My first job was in a school for the blind, at ten dollars a month plus room and board."

"You," Einstein broke in, "as an immigrant, at least have a job, which cannot always be said even of natives. When I received my diploma in Switzerland, I was unable to get a teaching position, and couldn't find enough students for tutoring to ward off hunger."

'You mean you — ?"

"Yes, in Zurich I often went hungry. No one knows how many doors I knocked at in search of a job."

"I always thought that your teachers were interested in your research on light. Or were they jealous?"

He looked up at the ceiling and opened his arms in a gesture of futility. "Oh, those professors! They had promised mc an assistant professorship after graduation, but later they forgot their promises."

"Perhaps it was anti-Semitism, or maybe they were afraid you'd outshine them."

"Possibly I wasn't smart enough for them," Einstein interjected. "You remember, I told you that I failed in school in Munich and I flunked the entrance examinations at Polytechnic Academy in Zurich." He added with a reminiscent smile, "Perhaps it was my old suit, which the professors had seen so often, and which made me unacceptable socially."

He shrugged his shoulders. "I would have taken any job, even that df a porter, for I didn't want to depend on my parents, who were no longer able to assist me financially. If it hadn't been for my friend Marcel Grossman, I might be peddling still. I owe my First job in Bern to him. He introduced me to the Director of the Patent Office. It was a petty job. I examined models of inventions, which didn't take long, so I was able to work on my theories. When the boss came in, I would shove my computations under a stack of papers."

"That reminds me of Heinrich Heine writing poems when he should have been balancing accounts at his uncle's bank. He was caught and fired; you weren't. And both of you secretly produced world-famous things."

A faraway look came into his eyes. "At least that simple job gave me the chance to catch a beam of light now and then. How I loved climbing the steep, musty stone stairs to my little room each day! There at my old desk, with its dusty files piled high, I decided that true scientific thought is not possible without faith in the inner harmony of our universe, and from this axiom I developed my theory of relativity."

There was a long pause, making me aware that the fate of immigrants was a closed topic. I pulled out the article I had written for the Christian Science Monitor, and pushed it across the table.

"Allow me to become your peace emissary," I said. "Here is a proposal for the organization of an international police force. It made a sensation at Harvard."

Without looking at it, Einstein pushed it back. Two dark, almost scornful eyes seemed to pierce me. "Now is not the time to speak of peace. Now is the time to speak of war—a war to crush dictatorship for good."

"Don't think for a moment, Dr. Einstein, that I'm not engaged in the war effort. Like yourself, I have relatives and friends still in Germany, and I lie awake nights worrying about them. There are terrible rumors about concentration camps, and thinking about my sister Gretel and her little daughter Ursula nearly drives me mad. And where are our many friends and coworkers who resisted the Nazis?"

After a silent pause, I continued, "Upon hearing of the rescue of the British soldiers at Dunkirk in June 1940, I wrote an anthem to awaken the Americans from their isolationism. Allow me to read it."

"Please do," said Einstein, as he leaned back in his chair.

FREE MEN GO FORTH

Free men, go forth to seize the shore
And onward surge—a living wall.
Move boldly toward the flames of war,
Drive evil to its fall.
Come touch the heavens with your hand
And vow to hold and never flee;
No man shall rise in any land
to force you to your knee.

Beyond the foe wields scourge and rope
And tears the night with rolling drum.
The hostage dies but not his hope—

Behold, the free men come!

Flow wave and wind until you meet
The terror, lashing seas and skies,
And fight on beaches, in the street,
Until all evil dies.

Go fight where snowdrifts build a cave,
Where deserts bum the air you breathe.
Your soul is greater than the grave
And needs no stone, no wreath.

0 life, cut short by tyrant's rope,
0 death at dawn with rolling drum.
The martyr dies but not his hope—
Behold, the free men come!

While I read him the poem, he turned his face away and gazed out the window at the morning sunshine.

"I composed the music, also," I explained, "and it has been sung by Harvard students on several occasions. I have the score with me, if you'd care to see it."

Einstein looked at me absently.

I tried once again to get Einstein to consider my proposal on the formation of an international police force. I pointed at the article. "This plan made a great impression on the round-table conference I organized for the Massachusetts branch of the League of Nations. It's too bad the Queen of Holland was hindered from coming."

"Oh, you tried to get her, a famous refugee?"

"And you, too," I interjected, "as well as former Reichskanzler Bruening, who is now teaching at Harvard."

"Then you tried to draw a full house with big names."

"What else could I do, my own name is nothing. Besides, you have committed yourself." I took a piece of paper out of my pocket and waved it in his face. "Here are your own words, your promise to help abolish all compulsory military service. I quote: 'Such a process might expand, and finally lead to an international police, the necessity for which would be gradually lessened as international security increased.'"

"Yes," said Einstein. "I mentioned that, but things have changed. For a long time my idea was to educate the people to resist compulsory military service, to force governments to look somewhere else for defense. A new type of mercenary army would thus be formed, similar to the Foreign Legion."

He shook his head dismally. "Alas, it was not to be." Then, in a sorrowful tone, "A dictator has come who thrives on the passive attitude of the masses. Education is replaced by propaganda. In the face of this new threat to humanity, it would be suicide to advocate nonresistance, whether pacifism or an international police force. Those panaceas would be sweet music for this dictator. By confusing the issues which vitally affect the life or death of a nation, they weaken the defense effort and thus become fifth columns."

"Are you then for what General Ludendorff once called total war?"

"He who has witnessed the atrocities of the Hitler Reich has no choice but to arm. Right now I would be the first to take up arms if I saw my dear ones being enslaved or killed and our civilization being wiped out."

There I was, sitting opposite Einstein, talking to him for the second time, but this time to a different Einstein. I was gripped by a feeling of awe for this man's courage in admitting his about-face so frankly and publicly. "Your change is a shock for many," I said. Einstein turned to me with a mystic's smile. "Yes, I know, I'm no longer an unconditional pacifist, but a realistic one."

The words you said the last time I saw you are still ringing in my ears: 'I would rather be cut to pieces than shoot someone on command.'"

He heaved a deep sigh. I'm fully aware that I made many enemies when I changed my mind about pacifism. I displeased the Quakers, of course, as well as the followers of Bertrand Russell, and of Gandhi. But principles are made for men and not men for principles."

I felt defeated. "Professor Einstein, I didn't come here to see you just for myself. We at Harvard have a defense group to which President Conant, the Philosopher Ralph Barton Perry, and historian Sidney Fey belong. I've made numerous speeches in Massachusetts exposing Hitler's propaganda. But we are also concerned with postwar problems and have founded a committee of Harvard students concerned with an international police force. On their behalf, I appeal to you to become a member of this committee."

Einstein raised his hand in protest.

"Become a member of this committee," I pleaded, "to help build up the idealism of young Americans." And in a more subdued voice, "You sign so many papers. Why not mine?"

"Yes, I sign so many papers! But if I have signed ten, I have rejected hundreds. I haven't signed a French manifesto against anti-Semitism in Germany; although I admired this noble effort, I felt it improper both as a Jew and as a German to sign a French protest."

"But Dr. Einstein, it was you who inspired me to think cosmically and work for one world. I am your cosmic missionary."

"Come on, little Willi," he chuckled, "don't use such big words."

Knowing I had lost the second round, I crumpled the article put it in my pocket. Einstein looked surprised: "I'm tired of having my name used as a billboard for every cause."

"But don't you crush the hopes of the youth who think of you as the cornerstone of the 'No more war' movement?"

"We must now concentrate on winning the war, and to speak at this time of an international police force means to scatter our fire. Let us be grateful that the democratic nations have created a united army. It is better to drive out the greatest evil, Nazism, with the lesser evil, armies, than to have Hitler triumph because we lack proper military force."

"Do you have any reservations about our alliance with the Russians and their conquest of the border nations?"

"Yes," he answered, "but their lust for conquest has more to do with the Russian Weltanschauung (world view) than with Communism. The Baltic States, with ice-free ports, are the door to Europe;

Peter the Great and Catherine weren't Communists, and their armies were as much a danger to western civilization then as Communist armies are now. Whether we call it colonialism, imperialism, capitalism or communism, Russia has never changed her national ideology. So let's be grateful that she's now fighting on our side against Hitler."

"You must, nevertheless, be aware," I said, "that a war with Russia is inevitable after this one, precisely because of her ideology." Einstein shook his head."But how, then," I asked, "can we fight Russian nationalism, or what might be called 'pan-Slavism'?"

"Certainly not through nationalism," he retorted laconically. "We must find an ideology that surpasses that of Russia."

"You mean, turn your left cheek after being hit on your right — Christian ideology?"

"Nonsense!" Einstein exploded.

"What else, then, can match Lenin's dictum: 'It is true as long as it profits the State; if it ceases to do so, it becomes a lie.'?"

There was a period of silence, as Einstein looked down at his desk. Then he spoke, "Henri Barbusse asked me to sign a pacifist appeal before Hitler became Reichskanzler, but I refused because of its praise for Soviet Russia. I had come to the conclusion that the leadership was being racked by personal struggles of power-hungry individuals, driven to use the foulest means to accomplish their purely selfish goals, yet the Russian people were being denied their worth as individuals and freedom of speech."

"Barbusse should have learned from the French Revolution," I said. "When the bourgeoisie had dragged the aristocracy to the guillotine, they had nothing else in mind but, *'Ote toi que je m'y mette.'* (Get up that I may sit down.) They took over the castles and played the aristocrats. Isn't that the same in Russia? I have discovered as a sociologist that members of a group subscribe to the lowest common denominator, because the group has no conscience, only goals. And when the group defeats the enemy and takes over their prerogatives, human nature unchecked by conscience repeats the same story of greed for power."

Einstein nodded; he was a good listener. After a pause he said, "The cosmic man must be restored, the whole man who is made in the image and likeness of the arch-force, which you may call God. This man thinks with his heart and not with party dogma. As I have explained before, there is an order in the universe —a cosmic order — and humans have the possibility of understanding these laws."

Einstein leaned back in his chair; so did I, putting my writing pad on my knees. He added, "I have no doubt that the Allies will win the war."

I smiled, "Oh, you are my prophet again."

"Prophet or not," he scratched his head, "what I say is more often felt through intuition than thought-through-intellect." He went on, "I hope the nations will learn their lesson, that there is no security in national armies. After the war we must renounce all national military defense and create an atmosphere of international trust; there can be no disarmament without security, and no security without an international court and, as you suggested, an international army to enforce law and order. But not now —after the war!" He stared at the ceiling; his voice became subdued.

If you were in my place," I asked, "what would you tell the student committee at Harvard?"

"I would tell them to have nothing to do with evil. We must begin at home, and with our neighbor. It's a daily task, not for Sunday only. There is much selfishness in our youth, and too little reverence for life. Buddha, Moses, Jesus, and Gandhi —their ethical instructions are better than mine."

"But they have never saved the world from war."

"I know," nodded Einstein. "The world simply won't live up to their teachings."

"But is there nothing else to tell them?" I persisted.

"Yes," he said, wagging his head slowly from side to side in deep thought. "I believe in the free exchange of scientific achievements. Science and art are the only effective messengers for peace. They tear down national barriers; they are far better assurances of international understanding than treaties. Students of art and science should realize that they are called upon to be citizens of the world."

I felt relieved at having a positive statement to take back. Leaning back in his chair, Einstein closed his eyes, and seemed to doze off. However, I had more on my agenda, and plunged undaunted into my last topic.

"Dr. Einstein, you said that matter is energy. Can't you dissolve energy into...."

A faint smile appeared on his sleepy face. "Back into matter? The word 'dissolve' is inappropriate." His glibness unnerved me. Still, I had come with a deep conviction that matter isn't real, and this time I was not unprepared. How cruel, I thought, to rouse him from his nap with quotations from Berkeley and Leibnitz, which I had collected in my notebook.

But then, his eyes still closed, Einstein asked, "Are you now better informed about light, or do you still think that when it is dark there isn't any?" The voice was calm, yet it struck with great force.

At last I stammered, "What a memory you have!"

Einstein sat up in his chair, clasped his hands, and rested his elbows on the table. "An educated man who does not know that light travels is not easily forgotten. Do you now understand more about my theory?"

"Not too much more," I answered, "but I like the stories about it. A physicist told your distinguished friend, Sir Arthur Eddington, 'You are one of the three men who understand the theory of relativity.' He blushed, whereupon the physicist said, 'Oh, Professor, you needn't feel embarrassed.' I'm not embarrassed,' replied Sir Arthur, 'I'm just wondering who the third man can be.'"

"Certainly it isn't you," Einstein said, "but don't be discouraged. Every year more and more students become familiar with the theory. It won't be too long before the principles of relativity become a part of basic teaching. Philosophy has already incorporated the rational concepts of physics."

"But how many people will really be able to understand it?"

Einstein smiled broadly. "It isn't important whether people understand this or that philosophical system. What they should understand is that they are endowed with a mind that has the power to unveil the mystery of life. This knowledge should make every man an individual thinker. If man becomes more aware of his dignity as a cosmic being than of his ego in the flesh, our world would then have peace."

Hastily pulling out my quotations, I said, "That's why I've come, to prove to you that matter, flesh included, is but a construction of the Five senses. This is from Berkeley:

All the choirs of heaven and furniture of earth —in a way, all those bodies that compose the bodies of the earth—have no substance without mind.

And here is Leibnitz:

I am able to perceive not only light, color, heat, and the like, but also motion, shape, and extension. However, all these are merely constructions of the mind.

Do you agree with this, Dr. Einstein?"

He looked a bit puzzled, and cleared his throat slightly. I rose saying, "Do these thinkers not prove the same thing, that matter does not exist?"

"No!" Einstein retorted. "You read into them your own thoughts." He glanced at me questioningly. "Suppose they did say it. Does that mean I should say it, too?"

"I believe you did say it, didn't you? For example, you said that magnetism or electricity may not be real."

"I probably said that our mental concepts of them might not be true —they can change with increasing knowledge." He took several deep breaths, and for a moment seemed to be baffled. Then, glancing out the window, "You must be able to distinguish between what is true and what is real. Magnetism and electricity do exist. Reality is determined by experience and contact, not by theories. This doesn't mean, however, that reality of the law of gravitation needs to be experienced by jumping off the Empire

State Building or that of electricity by touching a live wire."

My philosophical fervor got the better of me. I threw my note-book down and grasping the edge of the table with both hands, I leaned toward him. "Every object, whether it is as hard as this ta-ble, as sour as a crabapple, as hot as the sun outside, or as cold as ice, exists only in the mind. You must admit, therefore, that what you call energy, stars, and atoms do not exist, except as construc-tions of the conscious mind, and that they are conventional sym-bols shaped by the senses of man."

Einstein pushed his chair back and I hastily sat down. He smiled. "You said the sun is hot, didn't you, and that this table is hard?" He knocked on it several times for emphasis. "And I'm old and you're young? We aren't alone in saying this, you know; every man endowed with reason and five healthy senses would say the same thing. We all, more or less in the same way, say that a rose is red, smells like perfume, and feels like velvet. In other words, there is an objective reality which is conceived by the senses, and behind this objective reality are natural laws which are the privi-lege of the scientist to discover. Nature doesn't know chance, it operates on mathematical principles. As I have said so many times, God doesn't play dice with the world."

Fixing his eyes on mine with glee, Einstein continued, "When you carry a big suitcase, you would like someone else to carry it, right?"

"Of course!"

Einstein chuckled, "Of course, of course! You know as well as I then that matter is real, but with the difference that you allow your suitcase to be carried by others and I carry it myself."

"Do you consider the Bible a book that's been superseded, Dr. Einstein?"

"With regard to miracles?" he asked.

I nodded. With a smile that suggested sarcasm, Einstein said, "It seems that you believe in them."

"Certainly I do," I replied, taking up the challenge. "Although I would call the suspension of material laws divinely natural, rather than miraculous."

"God would not use supernatural force to retract what He created," Einstein retorted.

"One can only speak of natural forces when one has accepted the material world," I said. "Only the spiritual world exists."

Einstein seemed amused, so I baited the hook. "If God created the material world, then shouldn't we hold him responsible for all the miseries and woes of mankind?"

"We could," Einstein answered seriously, "if man hadn't been endowed with a free will. Every action of man is a function of his own or someone else's will. If so many people hadn't submitted their will to that of the Nazis, there would have been no concentration camps. You and I decided to leave and come here, because we felt Hitler was a reality."

He twisted a lock of white hair over his ear, laughed, and, to my astonishment, leaned across the table and slapped me in the face. "Did you feel that? Now can you say that matter isn't real?"

"Matter is real to my senses," I said, putting my fingers to my face, "but they aren't trustworthy. If Galileo or Copernicus had accepted what they saw, they would never have discovered the movements of the earth and planets."

Einstein bent over and faced me, and I instinctively flinched. "But their minds were never separated from their five senses." He lifted his finger. "Mark this: it is true that only the smallest part of what we perceive and think can be accepted as scientific truth. It is true that much of our labor is in vain, but that doesn't mean that our senses cannot be trusted. Perhaps they have become degenerate, perhaps they can be developed; but our senses are realities, just as much as what we hear, see, and feel, and they belong to the world of reality."

"But didn't you, yourself, doubt the existence of atoms when you state that matter can be dissolved into electromagnetic fields?"

Einstein became agitated. "I have never denied the existence of matter! Nor you, when your suitcase is heavy."

"But haven't you reduced the world to mental phenomena?" I

pressed.

"Electromagnetic fields are not of the mind." He became impatient. "Creation may be spiritual in origin, but that doesn't mean that everything created is spiritual. How can I explain such things to you?"

He pulled at his hair, apparently at a loss for the right words or frustrated by his inability to make me understand. "Let us accept the world as a mystery. Nature is neither solely material nor entirely spiritual. Man, too, is more than flesh and blood; otherwise, no religions would have been possible. Behind each cause is still another cause; the end or beginning of all causes has yet to be found. Yet only one thing must be remembered: there is no effect without a cause, and there is no lawlessness in creation."

"What do you call law?" I asked.

"Law is that which, under the same circumstances, acts and remains the same."

"But," I persisted, "you taught us that all is relative. Does that mean there are no absolute laws?"

"Yes. Nothing is absolute. Laws are relative to circumstances, and change comes without knowledge of them."

"Would you say, then, that God is mutable?"

Quick as a flash, Einstein retorted, "What do you mean by 'God'?"

"Couldn't I also call God the originator of natural laws?"

"If you wish, but remember this is not a personal God."

I rose from my chair, "I know you said once that most people fashion their God after their own image. How do you fashion your God?"

"To me," said Einstein, "God is a mystery, but a comprehensible mystery. I have nothing but awe when I observe the laws of nature. There are not laws without a lawgiver, but how does this lawgiver look? Certainly not like a man magnified. Maybe," Einstein nodded with a sad smile, "some centuries ago I would have been burned or hanged. Nonetheless, I would have been in good company. Giordano Bruno and the heretics through the ages were often people with deep religious feelings."

"It was Giordano Bruno," I added, "who said to his religious henchmen, 'It is with far greater fear that you pronounce this sentence to be burned, than I receive it."

Einstein said, "Bruno discovered that what man perceives of this world depends on his position in space and time. Dogmas, like scientific theories, rise and fall."

There was a silence. Einstein seemed to enjoy my religious struggle against his cool logic, and stared over my head into the air. I then told him how, as a refugee in Paris, I had once been on the brink of suicide, and was only saved when a friend who was a Christian Scientist emphasized that I, as a spiritual being, could not "die."

Einstein paused for a moment. "I remember Stefan Zweig once wrote something about spiritual healing."

"Do you believe me," I asked boldly, "when I say that I often have been healed by spiritual means?"

"That statement can't be verified." His manner became brisk. "It can only be proved false. Of course, I don't deny that thoughts influence the body."

Good, I thought, there's still hope. "Professor Einstein, if you witnessed a spectacular cure through mental power, when all the evidence pointed toward certain death, wouldn't you say to yourself, 'If another mind is a regenerating force, then mind governs matter'"

"I never denied that mind channels power," he returned. "It can be used in any way a man chooses, and the stronger man's faith in his mind, the more he can achieve. Without faith in one's ability, man is nothing."

"Faith can move mountains," I offered. Einstein remained silent. "I cured a woman of cancer by telling her to read the 23rd and 91st Psalms while I did the same at home. And I also healed Chaliapin, the Russian bass, when he had the grippe before a performance of Boris Gudonov in Paris. But I'm not so arrogant as to think that I did these things: it was the spirit. Of course, the people I prayed for had faith and their faith made them well in cooperation with my meditation."

"My mother," Einstein looked at the ceiling, "prayed for me, too, when I was sick, but *chacun a son gout* (each to his own taste). I'm interested in the natural miracles of the universe, not human miracles." He paused. "If I hadn't an absolute faith in the harmony of creation, I wouldn't have tried for thirty years to express it in a mathematical formula. It is only man's consciousness of what he does with his mind that elevates him above the animals, and enables him to become aware of himself and his relationship to the universe."

"Dr. Einstein, what do you think of meditation?" He looked puzzled for a moment. "I mean, in the sense that the yogis practice it?"

"I saw some of them when I was in India. I was very impressed by their serenity and selflessness." He smiled. "I was infected with it, too. I remember being unable to climb into a *ricksha* because I felt that the weight of my body would degrade the human being who was supposed to pull mc down the street. I wouldn't have minded a horse or donkey, but a human being...."

I pulled out some notes. "Once, in England, I was at a dinner with people highly trained in meditation, among them Professor Suzuki who asked me to ask you if spiritual vibrations and electricity have the same original cause or force."

"I believe," answered Einstein, "that energy is the basic force in creation. My friend Bergson calls it Man vital, the Hindus call it *prana*."

I told him how we had left an extra plate on the table for him, but he interrupted with a chuckle. "I don't remember that I was there, even in spirit. I have great respect for Professor Suzuki, however; and I think the highest goal for any educator is to open the minds of his students. We are so small, and the universe is so immense. To make a student conscious that he uses only about ten percent of his mental potential, that he is a part of this immensity and can comprehend it if he wants to; this is the duty of any teacher who's worth his salt."

The conversation strayed, so I interjected, "Jesus, in his teachings, certainly stressed man's capacity for knowledge. Wasn't it "knock and it shall be opened? He performed miracles of healing."

"Sometimes I think it would have been better if Jesus had never lived. No name was so abused for the sake of power!"

"That may be true," I said, "still, Jesus was the greatest Jew. Every Jew should be as proud of him as of an older brother. Of course," I added with some hesitancy, "I don't want to be facetious. Jesus wasn't as much human as divine."

He looked at me. "Well, the New Testament says so, but remember" —he lifted his finger—"that its earliest text, Paul's Letter to the Thessalonians, was first written about twenty years after his death, and some of the books not until the turn of the century. The philosophical and political currents of an age and, yes, just the passage of time itself, have a way of dressing up history. Indeed, in the second century there were many passages *hineingeheimnist* (slipped in secretly), especially the notions of Greek gods and the Word made flesh. I seriously doubt that Jesus himself said that he was God, for he was too much a Jew to violate that great commandment: Hear O Israel, the Eternal is our God and He is one!' and not two or three."

I added, "Jesus never taught anything other than what Moses and the prophets had taught."

Einstein didn't seem to listen, for he said all of a sudden, "Look what mankind has done to itself in the name of Jesus! Look what the Christians are doing and have been doing to the Jews for nearly 2000 years." Einstein rubbed one hand across his forehead. "One doesn't need to be a prophet to state what I fear for the Jews in Germany. My experience is that the moment man's mentality is bent on national goals and marching feet, it is corrupted and ready for even the most savage acts against humanity to justify its goals."

After a pause, Einstein continued, "Never in history has violence been so widespread as in Nazi Germany. The concentration camps make the actions of Ghengis Khan look like child's play. But what makes me shudder is that the Church is silent. One doesn't need to be a prophet to say, The Catholic Church will pay for this silence.' Dr. Hermanns, you will live to see that there is a moral law in the universe."

That reminds me," I told Einstein, "of the prophecy you made many years ago in Berlin—the one concerning a partnership between the Nazis and the Catholic Church. A Polish priest horrified by the Nazi invasion of Poland and the reckless execution of Jewish men, women, and children, got his Bishop's permission to address Pius XII. The priest traveled to Rome and related how Polish Jews were forced to dig their own mass graves, then lie in them, face down, naked, fifty at a time; after which the Nazi soldiers, perched on either side of the trench, sprayed them with machine gun fire. The Vatican told this priest that he must remain quiet since the Concordat with Hitler (who promised to check Communism) obliged the Church to tread softly."

"There are cosmic laws, Dr. Hermanns. They cannot be bribed by prayers or incense. What an insult to the principles of creation. But remember, that for God a thousand years is a day. This power maneuver of the Church, these Concordats through the centuries with worldly powers ... the Church has to pay for it. We live now in a scientific age and in a psychological age. You are a sociologist, aren't you?" I nodded. "You know what the *Herdenmenschen* (men of herd mentality) can do when they are organized and have a leader, especially if he is a spokesman for the Church. I do not say that the unspeakable crimes of the Church for 2000 years had always the blessings of the Vatican, but it vaccinated its believers with the idea: We have the true God, and the Jews have crucified Him. The Church sowed hate instead of love, though the Ten Commandments state: Thou shalt not kill."

Einstein looked through the window and seemed to mumble more to the trees than to me, "I believe that I have cosmic religious feelings. I never could grasp how one could satisfy these feelings by praying to limited objects. The tree outside is life, a statue is dead. The whole of nature is life, and life, as I observe it, rejects a God resembling man. I like to experience the universe as one harmonious whole. Every cell has life."

He turned to me and smiled, "Matter, too, has life; it is energy solidified. Our bodies are like prisons, and I look forward to be free, but I don't speculate on what will happen to me. I live here now, and my responsibility is in this world now. To march and kill and then think one gets absolution—no! God works with laws. That does not mean that love has no healing power, but I deal with natural laws. This is my work here on earth."

There was a long pause. My pen trembled in my hand as I tried to get all what Einstein said down on paper, but he was considerate and did not speak. His words caused me to grieve inside, and I said, "I have had a great longing to become a member of all religions and then unite them into one religion of love, but I was shocked when I learned of the Concordat which Hitler signed with the Roman Catholic Church on July 20, 1933. Pope Pius XI then asked God to bless the Reich, and this after Hitler instituted the boycott of Jewish shops with the declaration, 'I believe that I act today in unison with the Almighty Creator's intention. By fighting the Jews, I battle for the Lord." Einstein added, "Hitler also said, 'Conscience is a Jewish invention.ro "Dr. Einstein, I recently visited former Chancellor Bruening in Harvard. He, too, was a brother of your fate. He told me that from November 1933 he had been tracked by the Gestapo, and for weeks he had to sleep every night in another house. He crossed the border to Holland at the end of May 1934, very sick with a heart condition. He lamented the signing of the Concordat and told me that the Bishops were also ordered to swear allegiance to the National Socialist regime with the words: 'In the performance of my spiritual office and in my solicitude for the welfare of the German Reich, I shall take heed to avoid all detrimental acts which might bring peril to the Reich.'" Einstein nodded, "I'm not a Communist but I can well understand why they destroyed the Church in Russia. All the wrongs come home, as the proverb says. The Church will pay for its dealings with Hitler, and Germany, too."

He again paused; then said, "We Jews suffered for almost 2000 years. There was no means base enough to destroy the Jewish concept of God. How often were the Jews in the Middle Ages murdered en masse, accused of stealing the Host to desecrate it? And in England, how many Jews died because of the supposed miracles of the child Hugh of Lincoln, who is today still honored as a martyr? Poisoned wells, the plague, and every other conceivable catastrophe were blamed on the Jews, all because we did not accept Jesus as the Second Person of the Trinity, for we were among the first to proclaim that God is One. The Jews were ready to sacrifice even the Temple in Jerusalem rather than to allow the Emperor Caligula to place his image there to be worshiped."

His words were tinged with both pain and pride as he spoke of the past. Then, returning to the present he said, "With a few exceptions, the Roman Catholic Church has stressed the value of dogma and ritual, conveying the idea theirs is the only way to reach heaven. I don't need to go to church to hear if I'm good or bad; my heart tells me this. All that exists is based on a creative principle, and man's creative principle is his conscience."

"I agree," I said. "If people would make their conscience their Church, it would be better for them and the world."

Einstein continued, "I remember that you are a mystic and believe in a personal God, whereas I believe in the creative principle and the ordered regularity based on cosmic law. I don't like to implant in youth the Church's doctrine of a personal God, because that Church has behaved so inhumanely in the past 2000 years. The fear of punishment makes the people march. Consider the hate the Church manifested against the Jews and then against the Muslims, the Crusades with their crimes, the burning stakes of the Inquisition, the tacit consent of Hitler's actions while the Jews and the Poles dug their own graves and were slaughtered. And Hitler is said to have been an altar boy!

"The truly religious man has no fear of life and no fear of death — and certainly no blind faith; his faith must be in his conscience. Then he will have the intuition to observe and judge what happens around him. Then, he can acknowledge that everything unfolds true to strict natural law, sometimes with tremendous speed.

"I am therefore against all organized religion," he went on. "Too often in history, men have followed the cry of battle rather than the cry of truth."

"Isn't it only human to move along the line of least resistance?" I asked.

"Yes," Einstein replied vehemently, "It is indeed human, as proved by Cardinal Pacelli, who was behind the Concordat with Hitler. Since when can one make a pact with Christ and Satan at the same time? And he is now the Pope!" Leaning over the table, he continued in a bitter voice, "The moment I hear the word 'religion,' my hair stands on end. The Church has always sold itself to those in power, and agreed to any bargain in return for immunity. It would have been fine if the spirit of religion had guided the Church; instead, the Church determined the spirit of religion. Churchmen through the ages have fought political and institutional corruption very little, so long as their own sanctity and church property were preserved."

He extended his arms toward me, palms up, as though holding an open book, "Oh, Dr. Hermanns, the world needs new moral impulses which, Fm afraid, won't come from the churches, heavily compromised as they have been throughout the centuries. Perhaps those impulses must come from scientists in the tradition of Galileo, Kepler and Newton. In spite of failures and persecutions, these men devoted their lives to proving that the universe is a single entity, in which, I believe, a humanized God has no place. The genuine scientist is not moved by praise or blame, nor does he preach. He unveils the universe and people come eagerly, without being pushed, to behold a new revelation: the order, the harmony, the magnificence of creation! And as man becomes conscious of the stupendous laws that govern the universe in perfect harmony, he begins to realize how small he is. He sees the pettiness of human existence, with its ambitions and intrigues, its 1 am better than thou' creed. This is the beginning of cosmic religion within him; fellowship and human service become his moral code. And without such moral foundations," he continued wistfully, "we are hopelessly doomed."

"O, Dr. Einstein," pulling from my pocket a photograph, "I have a gift for you from the Empress Hermina."

Einstein, looking at the photograph, said, "And His Majesty with her. How be-medaled! Yet lost the First World War." He gave it back to me. "I'm not to be bribed."

I remarked, "What would Jesus say if he came back today? Perhaps what Lord Haldane said of you, 'Einstein is the greatest Jew after Jesus.5"

Einstein mumbled, "Leave me out of this, please."

"Didn't you say yourself that you admire nothing so much in creation as the mystical? Jesus is a Jew and we are Jews. After my mother died when I was seven, I was brought up by my Aunt Veronica, an Orthodox Jewess. Listen to what she taught me: 'Let the left hand not know what the right hand gives; Let the sun not go down before you have paid your debt because you know not whether your creditor has bread enough in his home to feed his children; One is more successful by praying to God than by trying to take from God.'"

"And this she taught me: 'No gold is more valuable than the gold of conscience.'"

Einstein nodded, "Yes, I agree. Nothing is so valuable as charity and love for our neighbor."

"Let us speak of mysticism, Dr. Einstein. After my Aunt Veronica passed away in 1929 on the threshold of the ensuing Nazi hate-storm, I would enter any open church door and proceed to meditate in front of another Jewish woman. The Mother of Jesus. For me, mysticism is as infinite as creation."

Einstein smiled, "Yes, the poet."

"But this love of mine," I continued, "is a part of the moral code you speak of. And isn't this moral code included in all religion?"

"Yes, but it's based more or less on fear; their God has human features. Let me put it this way: there are three phases of religious development. First, that of primitive man, who, looking at the stars and the ferocity of natural phenomena—fires, earthquakes, diseases — needed a mediator who had power to appease these incomprehensible forces. Over many generations, therefore, a caste of priests evolved who were rulers, healers, and politicians. But even these primitive tribes occasionally had scientific minds who tried to interpret nature's law of cause and effect; they thus gradually undermined the concept of the gods as angry and capricious beings."

Einstein chuckled as he noticed my hand trying to write down all he said and he asked, "Aren't you tired yet?"

He continued, "So, from the understanding of causal relations in creation evolved the concept of a God of justice. He became humanized and comprehensible; man could look to Him for guidance and consolation. But man still needed religious mediators—the clergy. And even here, on this highly moral and civilized level, the marks of primitive religions have not been effaced. Religious leaders have mixed their faith with politics and called for the destruction of other beliefs and of other races. Look through the ages at the array of scientists and philosophers who have been persecuted for their convictions."

"Yes," I remarked, "a few centuries ago you would have been burned at the stake."

His eyes twinkled. "I would have been in good company with those witches and heretics." Then his voice became solemn: "Men are now moving into the third phase of religious experience: cosmic religion. With his growing knowledge of the vastness of the universe and its trillions of stars, each one many times larger than our planet, stars whose light takes hundreds of years to reach our eyes, man must consider it an insult when he is told that his conduct should be motivated by fear of punishment or hope of reward. And it is just as much an insult to the God who created all these marvels, to be lowered to a human level. The true religious genius has always been endowed with this sense of cosmic religion, and was considered a heretic because he needed no dogma, no priestly caste, no humanized God. Some of the psalms and some Buddhist literature breathe this cosmic religion; so does the heathen Democritus, the Catholic St. Francis of Assisi, the Jew Spinoza. Believe me, what religious leaders were never able to do—to tear down the barriers between races and religion—the scientists of our age may do." He ran his hands through his hair.

This was illustrated," I said, "by British scientists, when they verified your theory of gravitation. The British had never given such a reception—even to their own King—as they did for you, a German, and this while the Battlefields were still not dry from the blood of soldiers killed by the Germans."

Einstein nodded and seemed pleased. "And certainly hatred for Germany hasn't deterred their Astronomical Society from its two expeditions to photograph the position of the stars during a total eclipse of the sun."

After a pause, he added, with a faraway look, The new missionaries who tear down the barriers of prejudice will no longer carry the cross. They will be the students of today and the scientists of tomorrow, with ideas as infinite as the universe. They will no longer read fairy tales about creation, but only the one authentic book written by creation itself. They will discover through mathematical computation the laws governing the universe."

"According to what you say, God must be the master mathematician." I was faintly ironic.

Einstein threw himself back in his chair. "I stand with the ancient Jews, who didn't dare write the name of God in their books, much less say it aloud. It's enough for me that the principles of creation are inherent in mathematics. And with pure thought I can grasp reality."

Einstein rose and so did I, but he continued, "I agree wholeheartedly, the Bible is the greatest book ever written. From beginning to end it makes man aware of his conscience. But the masses don't care for conscience. They care for *panes et circenses* (bread and games) and those who rule the masses want to please them to stay in power. So Hitler gave them *circenses* by burning books...."

"The Bible included," I interrupted.

He continued, "And also the Reichstag, the synagogues and, as rumor goes, they are now burning human beings."

There was a silence. I whispered, "Poor Poles and Russians."

Einstein added, "Poor Jews."

He sat down and I followed suit. He went on: "But for me this world, real as it may be, has never pulled me away from my duty on earth as I see it: to discover the laws governing the universe." He stopped and scanned my face with an indulgent smile. "I feel I have disappointed you."

"Not at all," I stammered, rather disarmed by his concern. I started to put my notes together but he remained seated without a sign of impatience, and said dryly, "Tell those who sent you that the law of conservation prohibits me from saying that matter can be dissolved into mind. Whether mass is transformed to atoms, electrons, or motion, it's still a reality, a manifestation of eternal energy—I mean indestructible energy. This oneness of creation, to my sense, is God."

"Yes," I said, "and this Jewish prayer, thousands of years old, Jesus learned from his mother: 'Hear O Israel, the Eternal is our God, the Eternal is One.'"

He rose from his chair, but I remained seated to write as he continued, "This concept of God will unite all nations, Russia included. A new age of peace will be inaugurated when all people profess a cosmic religion, when the youth have become laymen with scientific minds, and I'm sure the priestly class will submit, as they eventually submitted to Galileo and Kepler."

Einstein envisioned this humanistic spirit reforming the militarized schools as well. "The aim of education in a free society," he said, "must be the training of independently thinking individuals who, however, see in the service of the community their highest value. Man does not lack the intelligence to overcome the evils in society," he declared. "But what he does lack is the selfless, responsible dedication to the service of mankind. Religion teaches this selfless service from the pulpit, but the Concordat proves: 'Do as I say, not as I do.'"

We were both silent a moment. Einstein apparently wasn't offended by my attempt to convert him or my repeated challenges to his theory, because he smiled. "My theory of relativity has nothing to do with theology; it has never caused me to believe more or less in God. Do you imagine that Spinoza's philosophy was influenced by his job cutting diamonds? Man is what he thinks, not what he does. The basis of true thinking is intuition: this is what makes me abhor our present-day school system. They split each science into several categories; yet truth is only attained by a totality of experience. I was never attracted by specialization, I always wanted to know nature, creation itself. The mystery of life attracted me. My religion is to use my thinking faculties, as much as I can, to know what seems unknowable. Have you ever stopped to consider that reading books or gathering facts, has never led to any scientific discovery? Intuition is the prime factor in our achievements."

Leaning toward me, he lowered his voice and continued. "My concept about relativity had to do with my feelings rather than my intellect. I felt that the universe is never static; but I sensed, as the Greeks did, that 'All flows.' I think with intuition."

"Dr. Einstein, I have something more to discuss regarding intuition. You influenced me in 1930 on the way to the police station in Berlin..."

Einstein interrupted, "Can't you drop this from your memory!"

"No, I can't, because Spinoza taught me that we have eternal existence. The mind, he said, is eternal and moves this body for a while and then has other assignments, as it were. Our existence cannot be defined by time. This gave me a lot of insight, and my interest in Spinoza I owe to you."

Einstein chuckled, but nonetheless looked expectantly to how I would continue.

"From Spinoza I was led to Carl Jung, one of your friends— no, rather we have you as a friend. Jung said, and I know you admire him too, 'What happens in space and time proves an interdependence of objective events with the psychic states of man.'"

"I agree," said Einstein. "There is a pre-established harmony manifested in cosmic laws and related to our minds. You and I are Individuals; still we are held together by a pattern. We all have living experiences infinitely patterned."

"But connected," I added, "with former lives."

"*O, mein Gott!*" sighed Einstein.

"And I shall prove it to you. I was baffled to read that when our Jewish brother Jesus asked, 'Who do men say that I the son of man am?' he was answered, and this by believing Jews, John the Baptist, or perhaps Elijah, Jeremiah or another prophet.' You said once that God doesn't play dice. I interpret this as saying there is no chance happening, including my visit to you in 1930."

Einstein shrugged his shoulders and threw his hands up in despair.

"Yes," I smiled, "I'm a mystic, and you, too, because you love intuition. And I believe that the soul is eternal, or, if you wish, the mind is eternal and carries experiences from one plane of existence to another or from one time period to another—since we have bodies to fulfill a purpose, and they are involved with space and time. There is no chance happening in the universe. That I met that frantic writer at your door in Berlin, or that you received a compass and marveled at its workings, or that your uncle and later a student should take interest in you spells out the mystical relationship between the people we meet and our purpose. Perhaps I saved your life that day when I climbed up the stairs to meet that uncanny person before your door. And perhaps you saved my life, too. When I fled France, my entrance to England was barred in New Haven by Scotland Yard police, but a man watching the interrogation interrupted, This gentleman accompanies me to the funeral of the King.' I had tried to avoid this man on the ship crossing the channel, thinking he was a Gestapo agent, but he had insisted on talking with me. In the conversation the topic came around to you and the incident at your door. My experience with you endeared me to this man, who I later discovered was the Duke de Nemours. There are mysterious connections and mysterious powers, which help us to fulfill our spiritual purpose in life."

Einstein took a deep breath and said, "We both may have mystical connections, but my God appears as the physical world and stands for order, and if I can interpret this order or law, I haven't lived in vain."

"But I owe it to you," I said, "that I tried to interpret God as my conscience, filled with infinite spiritual threads, tying me to people

I may have met in my past lives, or to you, Liebermann and Hauptmann in this life, with good or evil. And my intuition, as your intuition, tells us which people are helpful to fulfill our purpose in this life and which not."

"I agree," said Einstein. "God gives us difficult problems, but all can be solved if we are true to our inner selves."

Einstein then cautioned me against automatic assumptions. "We need concepts to order our own existence —for example, this is good and that is evil, and there is God and there is the devil— and we need concepts to order the material world around us, like the law of gravity and time and space existing independently of matter. These concepts have been accepted for so long that no one questions their authority. We forget their human origin and think they represent indestructible truth."

"But we have to accept certain concepts a priori," I objected.

Einstein shook his head. "No scientific progress is possible if we accept a priori concepts without analyzing them and asking ourselves, on the strength of new experiences, if they are still justified." He had started to pace up and down; I perched on the edge of a chair and wrote furiously. "The world consists of real objects, and there are consistent laws underlying them. If we want to honor God, then let us use our reason and intellect to grasp these laws, which form the basis of a perfect mechanism. The concepts of space and time are many centuries old, but that didn't hinder me from questioning them."

"But," he went on, "let us not forget that principles, whether mathematical or moral, are free inventions of the human intellect. Therefore, if they have any value, they must be connected with our experience. The world is a physical reality which we perceive through our five senses, and there is nothing more destructive to science than having fixed notions about this physical reality. I don't say there's no heaven or hell, no God or devil, but these metaphysical attributes of reality have no interest for me. To me, it is heaven enough to serve mankind on earth."

"Dr. Einstein," I said, "you mentioned Francis of Assisi earlier; he was certainly one of mankind's greatest servants. I have given it a lot of thought. I'd like to become a Franciscan. I have already joined the Quakers and the Swedenborg Society, and study the Hindu scriptures—all to experience what you call 'cosmic religion.'"

"Why not?" Einstein interrupted. "I've read a prayer of St. Francis: 'It is better to understand than to be understood.' To me, cosmic religion means one humanity, one love, one peace."

"But you know," I continued, "even the Franciscans could not escape the sociological law: To keep or attract members, unite them according to a low common denominator. If Francis had lived longer, he would have been cast out of his own order. He was already treated by some brothers as belonging to the 'out-group.'"

"Yes," said Einstein, "no matter how idealistic and necessary a group is, each member must first be loyal to his conscience."

"Oh, Dr. Einstein, I just remembered an interesting story concerning conscience and Germany. As you have most certainly read in the newspapers last summer, eight Nazi spies were arraigned before the military court in Washington for being sent to the United States for sabotage. They had been trained to cripple railroads with explosives and to damage bridges, tunnels and major switching posts. They were to plant bombs in factories essential to the war effort and in department stores owned by Jews. Well, I went with the photos of the eight to Gordon Allport, the leading psychologist at Harvard, and told him that I would like to travel to Washington to speak to the Attorney General, Francis Biddle, about the accused and perhaps save them from the electric chair."

Einstein sat down. "The Germans can be killed or constrained, but they cannot be educated to a democratic way of thinking and acting within a foreseeable period. Behind the Nazi party stand the German people, who elected Hitler, after he had in his book and speeches made his intentions clear beyond the possibility of a misunderstanding. You are naive, Dr. Hermanns."

"Allport said the same to me, but nonetheless he wrote me a letter with an introduction to the Attorney General, but on handing it to me said, 'As a psychologist to a sociologist, once a group mind is formed the individual is encouraged to place his security in the group rather than in his conscience.'"

"I agree with Allport," Einstein interjected.

"Nevertheless, early last August, I traveled to Washington with the letter and was led to the seventh floor of the Justice department to Biddle's office. The walls bore the portraits of all the Chief Justices of the Supreme Court. I was received by Oscar Cox, the Assistant Attorney General. He was impressed by my intentions to found a world youth movement under the sponsorship of you, Dr. Einstein. Mrs. Roosevelt and the Quakers."

Einstein straightened in his seat, "Now I hear I'm a sponsor of your worthy if ambitious plans, Dr. Hermanns, but I unfortunately have little time to help you."

I continued with my story. "When I showed Cox my personal alibi in the form of my now translated poems War and Verdun, he said, 'I would like to do everything to help you.'"

"Open sesame! The magic word. I requested to see the accused Germans.

"He looked up, 'Why? What has that to do with this? If there is one among them who repents and says, "I am sorry that I signed up for this mission," then perhaps the hovering death sentence may be averted. And what a luster for our cause it could be, if we could train that man!"

"That is what I thought. If only one was there who knew that he had escaped the chair because of his repentance, how wonderful he would be in service to humanity by helping me to create cosmic-minded youth.

"Cox, young and athletic as he was, waved me aside, 'Hermanns, you are a dreamer. Reality looks different to these men. Every word they speak among themselves is recorded. And do you know, there is nothing but bickering, hate and disdain? Each blames the other for failure. You plead that they may have a conscience, but it certainly is not evident. They are Germans, trained in what and how to think. They will remain so until their bitter end.'

"All Cox's gratuitous hospitality, his wine and tidbits, did not comfort me. The next day I stood again before Professor Allport in his Harvard office and told him that my mission had failed. He didn't say 'I told you so;' instead, he invited me to lunch at the Harvard Club to soothe my feelings."

"He is a good man," said Einstein. "If I remember correctly, six of the Germans were executed in the electric chair and two received long sentences of confinement with heavy labor, a lighter punishment for their aid in the investigation and arrest of the others. It is a tragedy that men will die for a nation or cause that is evil. How deaf and strong-willed they must be not to hear the voice of conscience pleading for love of one's fellow man."

The striking of the clock reminded me that it was getting late. "Dr. Einstein," I said, "I feel guilty; perhaps someone is waiting for your luncheon."

He smiled and looked with such warmth into my eyes that I felt I had a fatherly friend.

"There are so many hungry people in this world; a little hunger makes for good company."

Einstein looked at me for a moment and while I extended my hand in farewell, he rose from behind his cluttered desk saying, "We both will live to see the destruction of Hitler. Nothing is left from Frederick the Great; nothing is left from Bismarck. Poor Germans who love to be ordered around and forget that it is more important to know the laws of the universe than the laws of the land! The cosmic laws cannot be bribed; they come home to us. As it is written, 'For God a thousand years is as one day.'"

"Speaking of victory, Dr. Einstein, do you plan to meet Churchill when he visits America? It's rumored that he will come to Harvard, too, but no one knows when."

Einstein announced, "No, I gladly leave the glory to Churchill. It will be his day. Yes," he looked out of the window, "I've met him before; he is an infinitely wise man. Churchill has prophetic insight." He smiled at me, "He's a little of a romantic, like some others I know."

"They told me in England what Churchill had said about you: 'At least I met one man who has conquered the world with his pen and not with the sword.'"5

Einstein took a piece of paper and tapped it, "That is true. I would like to finish developing the unified theory of gravitation."

Einstein, seeing me put my notes together, probably with a dejected face, patted me on the shoulder, "Cheer up, nothing is lost. You wrote once, of your experiences on the Battlefield of Verdun; why not use your fine gift of observation now to write about Churchill in Harvard?"

"Oh, Dr. Einstein," I reached into my inner pocket, "I have here a short story, which I forgot to give you. I met my first cosmic man after you, an old Negro woman in South Africa. She collects the discarded babies at the foot of Table Mountain and cares for them. Read this and you will be proud of me."

Einstein patted me on the shoulder and chuckled, "I'll read it now."

He sat down at his desk and in ten minutes he had finished the story. I could see that he was moved. He looked up at me, standing before his desk, and handed me back the manuscript. "Dr. Hermanns, continue with your writings. This is your best means of changing the heart of man."

I stammered, "I'm so happy to have been with you again. I will try to be worthy of you."

My head was throbbing as I tucked my notes into my pocket and left. As I crossed the lawn, I looked back at the window behind which Einstein stood, and felt nonetheless defeated. He hadn't agreed with anything I proposed, whether immigrants, international police force, or mind over matter. Yet, was I really defeated, when I left richer in thought than when I came?

I passed the boy who had been mowing grass that morning — now he lay sound asleep under a tree. I stopped and gazed at him. Einstein was right; the boy was real and wasn't merely a concept. His existence would continue whether I thought about him or not. His millions of ancestors were real, the present war was real. I continued on and contemplated writing a note of apology.

"No, not an apology," I said aloud. "I shall write about Churchill at Harvard and bring it to Einstein."

The Third Conversation

Introduction

In the five years which elapsed between the second and the third conversation, one monumental event for me was Churchill's visit to Harvard to receive his honorary doctorate. Since Einstein and Churchill had met when he was a refugee in London, I invited him to join me for the event, but he declined my offer. He asked me to write about Churchill's visit. Thus I had a reason to wear out my welcome once again, but the visit materialized five years later. During the war I was involved in lecturing about Hitler and his book, *Mein Kampf*. I worked also for a while in the Office of Strategic Services in the pay of a Major.

After the war had finished and I learned of the death toll in concentration camps which wiped out most of my family as well as more than two hundred of my friends and comrades, I became so depressed that I left Harvard on the advice of friends to get my healing in the quiet of a Christian Science sanatorium in San Francisco. There were no medicinal supervisors but Nature, with its more perfect healing power, of woods, birds and flowers, and the nearby ocean.

After my stay of six months in that island of beauty, I accepted a professorship for German at San Jose State College. And there it happened again—an outer event which was a sign of the principle of synchronicity at work. Robert Merritt, one of my students who was preparing for the ministry, introduced me to the Dutch evangelist, Corrie ten Boom, in a Protestant mission camp in Ontario, Canada. She told me that Father Coughlin, the Catholic radio priest in Michigan, was imitating Goebbels propaganda techniques and had chosen Einstein as his prey, claiming Einstein was a communist. When another guest at the camp, Reverend James, emphasized that upon returning to his home in Princeton he would make an appointment for me, I naturally embraced the idea. What an opportunity! I had nourished topics of discussion in my mind since Hitler's demise — world security and an international police force, Churchill at Harvard, and above all, how to help the Germans now, of whom Hitler had said almost with his dying breath, "Not a German stock of wheat is to feed the enemy, not a German mouth to give him information, not a German hand to offer him help. He is to find nothing but death, annihilation and hatred...'"(1)

Reference:
Langer, p. 237.

Third Conversation With Einstein

September 14, 1948

I was spending my vacation with my sister Hilda in New Rochelle, New York, when Reverend James telephoned that Einstein would receive us at the Institute for Advanced Studies. I asked Hilda to drive me down to visit Einstein, and her two sons and daughter came along, all equipped in true American fashion with cameras. Later, the Reverend James joined us at Princeton. My sister and niece left us to stroll about the beautiful grounds of the Institute.

The secretary couldn't hide her shock when she saw that I was invading Einstein's privacy with three strangers. Only after I introduced them did she appear reassured and led us through the long corridors to the last door.

When we entered, Einstein rose from his desk, which was covered with papers and books. Behind him was a huge blackboard filled with a hundred formulae and a warning to the cleaning woman: "Do not erase!" I presented my two nephews and the minister, and he made a gesture for us to sit down. But we discovered we were one chair short. The secretary, moreover, had left the door open. Did she mistrust us? We all remained standing.

I was confused. Certainly I would have preferred to come alone, but it would have been most ungrateful to exclude the Reverend James and my two nephews. Einstein didn't budge; he was as helpless as a child. Did he expect to conduct the interview on his feet? He looked at me with pitiful eyes.

I begged him to sit down. He answered with a helpless gesture, "But we do not have enough chairs."

I had important questions to ask, however, and thus was encouraged to say, "Be seated, please. My nephew Edgar is the youngest and doesn't mind sitting on the floor."

I pulled my chair up against Einstein's desk, concealing a small writing pad on my knees as much as I could. He turned his large brown eyes to me. I gave him a smile but he didn't return it, instead glancing first at the minister and then back at me. I began to feel uneasy. But there was no time for formalities; at any moment some other visitor might be announced by his zealous secretary, and I had to have some answers about Germany, a world government with Russia and the atom bomb.

"Dr. Einstein, don't you want to return to Germany?"

"I?" He looked bewildered and shook his head.

"Can't we be of more use where we are most needed?" I asked.

But now Einstein was all there. "I don't need to go to Germany to think better than I do here."

I said, "The Germans need us, to help them build a new Germany, a new world."

Einstein grinned. "We can build the new world here."

"Yes, a Christian world of peace," Reverend James joined in.

"Do you hear from Germany?" I asked.

"Yes, I get many letters every day, though I answer only a few. It is unfortunate, but many Germans seem not to have changed."

I said that I had heard there were no Nazis in Germany any more.

Reverend James immediately added, "They are ashamed of their past."

Einstein grinned again. "There were no anti-Semites in Germany either, but there must have been six million decent Jews; every German says that he knew at least one decent Jew."

"Or," I said, "perhaps there was only one decent Jew in Germany and sixty million Germans knew him. Perhaps it was you, Dr. Einstein."

Einstein shook his head, making the fringes of his long hair billow over his ears. "I don't count, please." As he reached for his pipe, my two nephews and the minister stood in unison, apparently to give Einstein a light. When Einstein began to light the pipe himself, the three sat down.

"Our churches," said the minister, "got busy immediately after the war. Prayer meetings are now held everywhere, and there is hardly a sermon in which the Germans aren't reminded of their guilt and shame."

"It seems to me," I spoke up, "that according to letters I receive, the sermons haven't thawed many hearts. I had to translate a letter of thanks to an American student for a package he had sent a bombed-out German family. One sentence read: "Yesterday morning when I looked out the window and saw the snow and ice, and we had no coal in the stove, I told my wife not to lose courage— the old German God is still alive.'"

"Yes, the German God is still alive," mumbled Einstein, "and you thought we should go back there!"

"Another letter stated, 'It's kind of you to send packages, but if only you Americans could arrange cheaper rates on the railroads for us to visit the graves of our sons in France, Italy and Holland. You have all the money now and can travel where you wish; we are poor, yet our sons fell for their Fatherland as your sons did for yours. '"

"But we Americans fought for a different reason," countered the minister, with the vocal approval of my nephews.

"I believe," said I, "that the words of Marshal Foch are the most appropriate epitaph for Hitler's soldiers:

> *He forgot that man cannot be God. He*
> *forgot that above the individual is the*
> *nation and above mankind is the moral law.*
> *He forgot that war is not the highest aim,*
> *for peace is above war."*

"If a German general instead of a French general had understood this, there might never have been a Blitzkrieg," grumbled Einstein as he twirled a lock of his hair. "I shall never go back." I scratched out topic number one on my pad.

"We might well console each other," I said. "We can't forget Auschwitz."

Einstein nodded. There was a silence. He must have grieved for Germany and the fate of his kin, I thought. His hair was now silver-white; his parchment-like skin was etched with deep lines. But his brown eyes remained unchanged, still fiery and childlike. He wore a light blue sweater with no shirt collar visible, as if the sweater were all he had on. His bare toes stuck out from wooden sandals, like those of a Franciscan friar.

The secretary looked in. Seeing us in conversation, she fetched a chair for Edgar; then closed the door. I gathered my courage. "Dr. Einstein, I have a long-overdue apology to make. Let me confess that on my last visit, I tried to push you on a religious bandwagon by interpreting your idea that matter can be changed into energy to mean that matter was unreal. Of course, thousands of others, who wanted to make you one of them, also misinterpreted you."

Einstein smiled, "I know, I know. Sometimes I do side with minorities; I don't mind signing my name to the petitions of some minority groups, even if I have no time to investigate fully their motives. I always like to help the underdog, because as a Jew, you see, I feel one with him. But no one can make me say that matter isn't real. I have great respect for matter. I wouldn't study the theory of gravitation if matter were fiction. Of course, we shall never exhaust the experience of our senses, but we can still use it to explain the material laws of our universe."

"That was my mistake: to believe that the healing of disease, which is a noble task, could be effected by mindpower alone. I thought that prayers could replace doctors; I now realize that matter exists and has its own laws. Even the greatest saints die of diseases. I'm sorry, Dr. Einstein, that I tried to interpret your theory to make it fit my former beliefs. Of course," and I put some force into my voice, "man should be made aware of his fourth dimension."

But Reverend James had raised his hand, as though he were in school, and Einstein acknowledged him with a nod.

"You said, Dr. Einstein, that space and time don't exist in reality. How true! Christ proved this almost two thousand years ago." Einstein, who was drawing on his pipe, immediately withdrew it from his mouth and scratched his head. I wondered what would come next.

The minister removed his glasses and leaned forward. "Jesus could walk through barred doors; he appeared to several persons at die same time: space did not exist for him. He was seen on one side of the lake, and the next moment he was on the other side. Time did not exist for him."

Einstein was still scratching his head. I said, "Jesus was endowed with divine powers, why shouldn't He do these things? He promised us that, endowed with the same spirit, we could do even greater things than He did. He showed us the path to the fourth dimension. You also showed us that path, Dr. Einstein."

Einstein shook his head, "Not to that kind of fourth dimension."

The minister leaned back in his chair and smiled happily. "You see, Dr. Einstein, all is relative."

Einstein looked somewhat puzzled. "Yes, wise men have known this all along."

"Christ is more than a wise man," continued the minister. "The laws of physics did not exist for him."

Einstein retreated into his chair with a faint smile and hooked the pipe back between his lips. Watching the minister's pale, fervent face, it occurred to me that Corrie ten Boom's crusading angel was among us.

"One doesn't have to be a Christian to hold such beliefs," I interrupted. "In a conversation with Dr. Brahmachari, an Indian monk, I learned that in his Tibetan monastery the monks can walk through walls; and the sick, whether leprous or cancer-stricken, are healed miles away, so strong is the spiritual power emanating from there."

Einstein listened with kind indulgence. Then, addressing the minister, he said, "It suffices me to know the physical laws governing the relativity of time and space. I was not a student of Jesus, I'm sorry to say, but of Euclid, Leibniz, Gauss and Riemann." And, looking at me, he continued, "It is sheer nonsense to deny natural laws. They reveal such intelligence, that any human logic falters in comparison."

"The existence of matter is also necessary from the religious standpoint," said Reverend James, "for without suffering there is no purification and no coming closer to God."

Einstein threw him a surprised glance. "It is your privilege as a minister to justify the existence of matter from a religious standpoint. However, man should be proud of his ability to constantly check and recheck what one may call the synthesis a priori. He will discover that many concepts which generation after generation have considered absolutely true have limited or no validity. This goes for the Church, too."

"What is life on this earth anyway?" asked the minister. "The psalmist says, 'As for man, his days are as the grass; as a flower of the field, so he flourishes. For the wind passes over it and it is gone; the place thereof shall know it no more.'"

Einstein looked at him good-naturedly and then looked at me. I tried now to get the conversation back to my notes. "Dr. Einstein, you have often spoken about the degradation of science by nationalism. It is interesting to hear what the educators in Germany said about your relativity theory in Hitler's Third Reich. There is Professor Miller of the Aachen College of Technology who wrote that you tried to transform the living world,'.. .born from a Mother Earth and bound up with blood, bewitching it into a spectral abstraction in which all individual differences in people and nations, and all inner limits of the races, are lost in unreality, and in which only an unsubstantial diversity of geometric dimensions survives which produces all events out of the compulsion of its godless objection to laws.' Professor Tomascheck, Director of the Institute of Physics at Dresden wrote, 'Modern physics is an instrument of world Jewry for the destruction of Nordic science. True, physics is the creation of the German spirit...in fact, all European science is the fruit of Aryan, or better, German thought.' An old acquaintance of yours, Professor Stark, made known that "the founders of research in physics, and the great discoverers from Galileo to Newton to the physical pioneers of our time, belong almost exclusively to the Aryan, predominately the Nordic race.' And then there is Professor Lenard, whose diatribes against you in the Berlin Philharmony you witnessed, and who wrote in his four-volume Deutsche Physik: 'One may summarize all Jewish science by calling to mind the probably pure-minded Jew, Albert Einstein. His theories of relativity seek to revolutionize and dominate the whole of physics. In fact these theories are now down and out. They were never even intended to be true.' How must the Nazis have been pleased when they read such statements from this Nobel laureate: 'Science, like every other human product is racial and conditioned by blood.'"

Reverend James lifted his Bible and said, "This is the history of mankind, the only one worth knowing."

Einstein turned his head to face me, "Reverend James is closer to the truth. Those allegations of German scientists do not interest me."

"One of the biggest disappointments in your life," I continued, "was probably your experience with the Prussian Academy of Science." I then explained to my nephews how the Academy had turned Nazi and slandered Einstein, yet had the temerity to ask him to prove his loyalty by speaking well of the German people while abroad. But they got an unforgettable answer. "You said," and I turned to Einstein, "that to put a good word in for the German people would be a denial of the concept of justice and liberty, which you had fought for all your life; that it would be a barbarization of manners and the destruction of the values of culture. Nonetheless, Planck tried to intervene against Hitler on your behalf."

"Planck was a great scientist," said Einstein. "His discoveries were the basis for modern research on molecules and atoms. He was also a great man. I admired him deeply, even though I was never close to him. There was something chilling about him. He came from a family of Prussian officers and State officials."

"One might have thought that some members of the Academy would have resigned," I commented.

"I don't think so," replied Einstein. "Most were Germans before they were scientists; their loyalty to the State hindered their obligation to humanity. I say this without reserve, because I had an unforgettable experience with Prussianism," he continued evenly.

"When I heard of the invasion of Belgium by von Kluck's army during the First World War, and the slaughter in its wake, I couldn't help protesting to the Academy. Planck was there, Nernst, Roentgen and Haber, too."

"All Nobel Prize winners," I interjected.

"They all sat there," Einstein continued, "as if they didn't understand me. There is nothing I hate more than the violation of the internationalism of science, and when most of the members of the Academy signed that evil manifesto supposedly proving Germany to be the spearhead of modern civilization, George Nicolai, Wilhelm Foerster, and I drew up an opposing document, demanding peace without conquest and a united Europe. Nothing made me so sad as when I learned that my friend Haber, along with Nernst, had become a major in the German army and accepted an assignment to study poisonous gases. Haber once told me about the deadly gas bombs that Germany would soon use to bring the war to a victorious end."

"Yes, I got a taste of that gas," I recalled. "When we conquered the bunker at Thiaumont, we found hundreds of gassed Frenchmen. Unable to carry them out for burial, we shoveled earth down and buried them where they lay. Then we occupied the forts and had to sleep and live on top of them."

Edgar in his youthful vigor interrupted, "Who back then realized the importance of your discovery?"

"Max Planck," I said.

Einstein slowly shook his head. "Max Planck saw the importance of the theory, but shied away from publicizing it; he thought people couldn't grasp it. It is very difficult to pull people out of their traditional ruts, away from thinking that light travels in a direct line, for instance, or that time is absolute. Time is a concept of the human mind."

"You're right, Dr. Einstein." The minister brightened. "The seven days of the creation of the universe are not the seven days of our calendar."

Einstein smiled and nodded. "Of course," he turned to me,

"Max Planck began to correspond with me immediately."

"How long was it before you were noticed?" my nephew Herbert asked.

"Two years later. I was offered a position as an unpaid lecturer at Zurich University, but I kept my job in the Patent Office at Bern. Then Hermann Minkowski, my old teacher and friend, popularized my relativity theory by putting all my work into one sentence: 'From now on, space in itself and time in itself must yield to a union of both.'"

"I once read," I recalled, "that he lamented on his deathbed: What a pity that I have to die in the era of the development of relativity.'" Then I went on: "Dr. Einstein, I read that when you were offered the assistant professorship of Theoretical Physics at Zurich, you asked them instead to promote your friend Adler, who had been there for some time, because you wanted to remain in the patent office. How unselfish can a man be?"

Einstein frowned, "I was no success anyway," he said after a pause. "There were only a few people at my lectures, and one supposedly came to warm his feet because there was no stove at home."

"Well, when did your real breakthrough come?" Edgar asked.

"It was in the winter of 1913. Planck and Nernst came down from Berlin to offer me a position with the Prussian Academy of Sciences."

"It was well-known in the city," I said, "that you remarked, 'Now these Berlin people are speculating with me as if I were a prize hen. But I don't know if I can still lay eggs.'" We all laughed, and Einstein smiled down at the floor. "I remember how you were vilified by some students in the Humboldt Club when you were going to lecture abroad. I called you a worthy ambassador of science, and someone cut me short: 'As naturally as the liver excretes gall, so the brain excretes thoughts—and a German brain, German thoughts. Einstein has no message for the German people.' Some American and French students left the club in protest."

Einstein looked up. "Spinoza declined honors from the powerful on this earth and remained a lens grinder. If I should be born again, I will become a cobbler and do my thinking in peace. Oh, these universities with their power-crazed intellectualism of 'I am greater than the rest!' It's no wonder that scientists become instruments of tyranny."

After a pause, Herbert took the opportunity to ask Einstein how one could best explain his two theories of relativity.

Einstein leaned back in his chair and, gazing at the ceiling, said that for centuries people thought that if matter didn't exist, space and time still would. But the theory of relativity teaches that if matter disappears from the universe, then space and time disappear, too. The theory also teaches that energy causes movement; and that since mass is movement, the more speed increases, the more mass increases. Hence, the mass of a moving body is greater than the mass of a stationary body. With the equation $E = mc2$, we can calculate exactly how much energy we get if we change matter into concentrated energy."

He continued, "The general theory of relativity states that the laws of nature are the same for all systems, regardless of the state of motion, while the special theory of relativity states that the laws of nature are the same for all systems moving uniformly relative to each other."

"You made a powerful break with tradition," Herbert said.

"Yes, my theories did away with the Newtonian concept that the balance of gravitation and inertia was merely an accident of nature or just a force. I never believed that the earth could attract an object, or that this gravitational force could be the law that keeps the universe in shape; these were the illusions of mechanical concepts of nature. My law of gravitation contains nothing about force. Celestial bodies simply follow a path through gravitational fields according to the law of least resistance, as the marbles boys shoot follow the rise and fall on the surface they play on. The Newtonian theory of action at a distance is not correct. Each celestial body creates a magnetic field which is a physical reality and has a definite structure. Thus everything in the universe is movement. As the fish in the ocean agitate the water, so the galaxies agitate space and time in the universe."

"Is it true," I asked, "that a kilogram of coal changed to pure energy would produce 25,000,000 kilowatt-hours of electricity, which is as much as all the generating plants in the United States can produce in two months?"

Einstein nodded. "To simplify the concept of relativity, I always use the following example: if you sit with a girl on a garden bench and the moon is shining, then for you the hour will be a minute. However, if you sit on a hot stove, the minute will be an hour."

Einstein looked at some papers on his desk and I rose to ask if he were writing his autobiography. "Friends constantly urge me to do so, but I dropped the idea, for there is nothing worth writing down. My life hasn't been eventful enough."

"But, Dr. Einstein, you discovered laws of relativity which apply everywhere in the universe! That's not uneventful? And space telling matter how to move or matter telling space how to curve is to leap into another world, that before you was unknown."

He smiled and shrugged his shoulders, "Intuition and the opportunity to be at the right place at the right time. I couldn't have come to my conclusions without the discoveries before me of great scientists."

My nephew Edgar asked whether the Prussian Academy would invite him back now.

"I'm not interested," said Einstein.

"I guess you're glad to be an American now," said Reverend James.

"Yes, I am glad. I admire the Americans. In contrast to the stiffness of Europeans, they have childlike qualities; they are optimistic and don't harbor grudges. But most fascinating of all is the responsibility an American assumes for his community. Someone brought up in Germany has never learned the fundamental democratic principle: 'I'm as good as you, and you're as good as I.' For Europeans, this common denominator of equality is sometimes startling."

"Startling indeed," I added. "One can't imagine German students singing, 'The bright boys, they all study math/ And Albie Einstein points the path/ Although he seldom takes the air/ We wish to God he'd cut his hair."* Einstein grinned.

"Where would you prefer to live," I asked, "in Israel or in the United States?"

"I'm too old to be a pioneer," Einstein responded.

"But you are a Zionist, at least at heart."

"But not in the sense of narrow nationalism. I believe in re-building the Holy Land to give the persecuted a haven of peace and security, and a spiritual center in a world which, after all, began with our prophets." Einstein added that it might have been better if the Jews hadn't chosen Palestine for their home, but Uganda instead. "It's of course not ideal, but at least they would have had room. I believe it was the Bible that determined their choice —a nationalistic idea, and I'm against nationalistic ideas. I've seen too much of that with the Kaiser and Hitler."

"How can we solve the Jewish problem?" I asked.

Reverend James instantly answered, "Man must be born again, that is the only way."

Einstein smiled. "Whether born or reborn, I think the Jewish problem can be solved through Zionism, less as a political entity, however, than as a cultural one."

"But is that culture dependent on having a State?" I asked.

"Unfortunately, Hitler has made that a necessity."

"The Bible predicted the return of the Jews," said the minister.

Einstein smiled and I had to smile, too. Reverend James added, "The Jews had clever prophets —they even foresaw Hitler."

Blond, stocky Herbert asked whether the survival of the Jews is important.

"Yes," Einstein replied, "because Judaism is not so much a creed as it is an ethical code that sanctifies life. Walter Rathenau once said, 'When a Jew says he hunts for the sake of amusement, he lies.' The sanctity of life is so ingrained in a Jew that he hunts only when starving."

"We shouldn't be concerned with this life," said the minister. Einstein straightened himself. "But I am concerned with this life. The God Spinoza revered is my God, too: I meet Him every day in the harmonious laws which govern the universe. My religion is cosmic, and my God is too universal to concern himself with the intentions of every human being. I do not accept a religion of fear; my God will not hold me responsible for the actions that necessity imposes. My God speaks to me through his laws. Shouldn't we do good for the sake of doing good, and not because we fear punishment or hope for reward in a life to come?"

"It says here we should fear God," said the minister, taking out his Bible.

Einstein looked at me quizzically, and I felt my pad slip to the floor. While the minister was picking it up, I abruptly said, "A beautiful phrase, sanctity of life. It reminds me of that statue in the East Prussian forest which announces in gilded letters:

On this spot His Majesty, the German Kaiser Wilhelm II, shot his 40,000th creature, a golden pheasant-cock.

There was no sanctity of life at the Kaiser's court."

"What puzzles me," said Herbert, "is that the leading German generals knew Hitler was losing the war, yet kept on fighting anyway.

No sanctity of life."

Reverend James had found his passage. Bending earne toward Einstein, he read,

Yet will I leave a remnant, that ye may
have some that shall escape the sword among
the nations, when ye shall be scattered through
the centuries.

He assured us that the Jews of today could escape Armageddon because 'Blessed are they who are called unto the marriage supper of the Lamb. '"

Einstein listened patiently, and then, with faint sarcasm, said, "It's so puzzling to me that the God people want to convert you to is always good, while the God they convert you from is always bad. The Jewish God is a cruel God in the eyes of the Christians, and the Christian God a God of love; yet so much more blood was shed in the name of the Christian God. Let these apostles ask themselves: Why for centuries has my Church inflicted sufferings on the Jews?' If they did this, they would convert themselves first, and rewrite their religion and history books."

"That reminds me," I said, "of how Luther, who himself developed a hatred of the Jews, once admitted: 'First we beat them lame, then curse them when they limp.'"

"I feel the shame, too," replied the minister. "Dr. Einstein, we call this a Christian world, and yet must witness how it more and more becomes un-Christian. And why? Man loves sin more than purity. The world ignores what still holds true: Whoever drinks of this water I give shall not be thirsty for eternity.'"

"I think I am a religious man," said Einstein, "and no one will convince me that the world can survive without Jewish-Christian ethics. But I am very careful about accepting claims of authority. There is a mystical drive in man to learn about his own existence. And how can he achieve this? Galileo showed the way by creating a system of thought that binds together observed facts. I believe that the dignity of man depends not on his membership in a church, but on his scrutinizing mind, his confidence in his intellect, his figuring things out for himself, and above all his respect for the laws of creation."

"Jesus," emphasized the minister, "enhances man's dignity by adding spiritual values to it."

"So did Moses, Isaiah, Jeremiah and Buddha," countered Einstein.

"Christ was God's incarnation," said Reverend James.

Einstein leaned back in his chair. "In school I learned that every man was created in God's image."

"But that doesn't save man," the minister insisted. "It is Christ's blood that saves."

Einstein shrugged his shoulders. "I believe in one thing— that only a life lived for others is a life worth living."

The whole conversation went so rapidly that I hardly dared breathe for fear I might miss a key word. There sat Einstein, rocking so far back that once or twice he would surely have fallen if it hadn't been for his firm grip on the desk. He looked baggy and shapeless, while a few feet away, erect in his black suit, the thin, passionate Reverend James, glowing with enthusiasm, beamed the saving message at the serene scientist. Edgar and Herbert were sitting tensely on the edge of their chairs.

Wanting to avoid a storm, I interrupted. "Dr. Einstein, you stated that concepts held to be absolutely true may prove to have only limited validity. That reminds me of how some Protestant theologians maintain that Jesus never claimed any other title than 'Rabbi.' After his death, however, the founders of the new religion called him 'Son of God,' which was the prerogative of the Caesars, and 'Savior' that of the Greek kings."

"The Jews took their first commandment seriously," said Einstein. "To my knowledge they addressed Jesus as 'Rabbi' and perhaps thought of him as a prophet."

"They may also have given him the title 'Messiah'," I added, "thinking that he would help them get rid of the Romans."

"A lot of myth and legends, no doubt. Why not?" He glanced at me and chuckled. "There will be so many inventions written about me after my death; at least I'm in good company."

The minister tapped his Bible. "This is no myth! It is prophesied that when the Jews return to their homeland, the time of fulfillment is near."

"Oh, yes," said Einstein with a smile. "The newspapers report that some sects have sold all they have, they expect the end of the earth and will go straight to heaven without baggage." He became serious. Title or not, Jesus was born a Jew and remained a Jew, which some people like to forget. Half the stories in the gospels were added after the original accounts and many of them were handed down by hearsay, mouth-to-mouth."

"But the New Testament, like the Old, is inspired," protested the minister. "And my Church does teach that Jesus was a Jew, though a special one. I have two small children, and my wife and I tell them in a simple way of Jesus' birth to a Jewish family. And whenever the opportunity arises, I point out the mystery of the Jews—the mystery of their suffering, and that they have been and forever are the chosen people. We also believe that they will recognize Jesus on His second coming as their Messiah."

"Yes," mused Einstein, "Frederick the Great, that great atheist, must have known this mystery of Jewish suffering, too. When Voltaire told him there were no miracles, the king pointed out the existence of the Jews in spite of all their persecution. As for the Messiah, it's all speculation and certainly not worth to Battle over."

Seeing Einstein's good humor, I wanted to make him aware that I had no stake in the dialogue between him and the minister, "The Church can't deny she was the primary source of this persecution. During the Middle Ages she introduced the yellow star and the ghetto, and forced Jews from the country into the cities, where they could make a living only by lending money."

"The Popes did that," retorted Reverend James. "I'm a Protestant and I know one thing: we must do God's work, not man's." Then, turning to Einstein, "The Holy Spirit tells us how to do God's work."

Einstein clasped his hands and leaned forward. "If you mean that your conscience tells you what to do, then I agree." After a short pause, "But conscience, of course, can be manipulated. Many of the Germans who chanted 'Heil Hitler' were eager churchgoers. I agree with you, intellect never has saved the world. If we want to improve the world we cannot do it with scientific knowledge but with ideals. Confucius, Buddha, Jesus and Gandhi have done more for humanity than science has done. We must begin with the heart of man—with his conscience — and the values of conscience can only be manifested by selfless service to mankind. In this respect, I feel that the Churches have much guilt. She has always allied her-self with those who rule, who have political power, and more often than not, at the expense of peace and humanity as a whole."

"How true," I said, "a Catholic prelate in San Jose told me not long ago that when a friend of his in Rome, himself a high church dignitary, had asked Pius XII to help alleviate the starvation in China, he answered, 'I have first to think of my four hundred mil-lion Catholics.'"

Einstein shook his head, "But humanity is one and undivided."

"I often wonder," I said, "that even as early as 1933, Hitler dared to make a boycott on April 1 against the Jews without a pro-test from any leading churchman. He announced, 1 believe that I act today in unison with the Almighty Creator's intention: by fight-ing the Jews, I do battle for the Lord.'"

Reverend James snapped, "This God was, of course, the Catholic God, but not the God of Jesus Christ."

"How guilty the Church is," added Einstein, "we can learn from the heathen Roman philosopher, Seneca. He wrote, 'He who does not prevent a crime when he can, encourages it."

"We heard in the League for Human Rights that Hindenburg protested against Hitler's anti-Semitism," I said, "especially when directed against Jewish war veterans. He wrote Hitler a letter, say-ing that if they were worthy of fighting and bleeding for Germany, they were also worthy to continue serving their country in their own profession. The League also had its secret channels in Italy and learned that even Mussolini had protested Hitler's virulent at-tack against the Jews."

"Anyway," said the minister, "Hitler's Catholic God protected him with the Concordat. And I must admit that it has been estab-lished beyond a doubt that Rome knew about the concentration camps by 1942."

Einstein mumbled, "What happened to the Holy Spirit?"

"Dr. Einstein, be assured that my prayers include myself among the guilty. But I'm a missionary; I want to save souls for God, who has been so deeply offended. We are on the threshold of

Armageddon." He again opened the Bible.

I admired the noble bearing and intentions of the young minister, who wanted to wrestle with Einstein's soul as the angel once wrestled with Jacob's. I shifted the conversation. "Aren't religion and science contradictory? One is based on faith and the other on proof."

"Religion and science go together," said Einstein. "As I've said before, science without religion is lame and religion without science is blind. They are interdependent and have a common goal—the search for truth. Hence it is absurd for religion to proscribe Galileo or Darwin or other scientists." Einstein grinned, "And it is equally absurd when scientists say that there is no God. The real scientist has faith, which does not mean," he looked at the minister, "that he must subscribe to a creed. Without religion there is no charity. The soul given to each of us is moved by the same living spirit that moves the universe."

"Ah," said Reverend James, "you speak of the soul. Then you believe in a life after death."

"I believe," answered Einstein, "that we don't need to worry about what happens after this life, as long as we do our duty here—to love and to serve."

"You have every reason to be proud of being a Jew," the minister assured him, "for the Jews gave the world the Bible. The only thing I regret is that you don't heed the New Testament, which proves that Jesus was Christ. The Jews have no respect for miracles."

"You are mistaken," replied Einstein. "I have marveled often in my life. As a boy of twelve I already marveled at Euclid's geometry. Of course, what I thought was superhuman or miraculous was soon nothing but logical thinking added to experience. Still, I marvel every day, and it is my faith in the order of creation which makes me marvel. When I do so, however, I use logical thinking to find out why."

"Don't you agree, Dr. Einstein," I asked, "that faith is a mystery? It can't be comprehended by reason, but it does not contradict reason."

"I'm not interested in metaphysical speculations separated from sense experience," said Einstein. He paused, looking at each of us carefully; then he turned his attention to his sandals and gazed at his toes.

"Have you no faith in a life to come?" asked the minister.

"No, I have faith in the universe, for it is rational. Law underlies each happening. And I have faith in my purpose here on earth. I have faith in my intuition, the language of my conscience, but I have no faith in speculation about Heaven and Hell. I'm concerned with this time—here and now."

Memories started to flood my mind, and I felt I must share them with the others. "Dr. Einstein, you have always maintained that life is mysterious. Indeed, I myself have experienced that there is more to me than what I behold in the mirror. Once I was in utter despair about my sister Gretel, who had been carried away with her daughter to the Theresienstadt concentration camp. And many, including one close relative, thought it was a model camp supervised by the Red Cross. But I felt that it was a staging for Nazi propaganda."

Einstein interrupted, "Your relatives used their intellects, while you used your feeling. One never goes wrong following his feeling." Turning to the minister, he added, "I don't mean emotions, I mean feeling, for feeling and intuition are one."

"Exactly!" I added, "One night I was in such despair about the fate of Gretel that I began to pray: 'God, you heard me on the Battle-field of Verdun and saved me. Can you not help Gretel?' I must have fallen asleep, when suddenly I felt I had left my body, passed through the air and landed before a camp. I went through the barbed-wire, entered one of the barracks and stood before a woman on the floor with graying hair, thin to the bone; only her large dark eyes were the same when she raised her head to look at me. She said, 'I am Sarah Loewenhardt.' 'Gretel,' I called out, 'there is a God; where is your daughter?' She pointed to a little shadow; lying in a niche was Ursula. I then lifted my finger over my head, and mumbled, 'God.' And with this I awoke in my bed at New Rochelle, where I was visiting my sister Hilda. Not long ago I met a woman who was in Auschwitz. When I described to her my vision she said, 'Your vision sounds exactly like Auschwitz. We women had only one name, "Sarah," and the men, "Isaac," with our family name, for other first names were reserved for those with pure Aryan blood. In the moment someone approached us, we had to say our name, before it was asked. Your sister thought you were one of the guards.' I've written a poem about this experience. Allow me to read it:

GRETEL

I came to comfort you; the
stars told of your plight and
gave my body wings to fly
here through the night

Come, lay your cheek on mine,
our tears together flow.
0, solitude of death,
how long its shadows grow.

Come, put your arms around mine,
My heart shall feel your heart,
until its slowing beat is ready to depart.

My soul I give to you
into your earthy bed,
that you may rise again,
and I lie there instead.

Einstein looked at me, took off his glasses and, deeply moved, said, "I lost most of my family."

There was a silence, and then the minister said, "Yes, there is a God. Most people don't remember that there is more to man than meets the eye."

"I'm still hung up on 'sanctity of life,'" said Herbert. "During the war I served as a lieutenant in Alaska. We learned how to take care of our huskies; they were vicious, but we treated them well. We felt we were one team. But the German generals weren't concerned much even about their own people. In the Nurnberg trials Field-marshal von Blomberg was supposed to have said that an officer's only concern was promotion and medals."

Edgar, also a former G.I., addressed the minister. "I read that the generals always hoped Hitler would win the war, and so didn't want to know about the massacres in Poland or the concentration camps. They wanted to stand great before history but not before humanity. And weren't they Christians?"

I added, "Yes, and remember General von Froeben, Frederic the Great's friend, who helped us with our war for independence? In a letter to his wife, he wrote he was amazed that the American soldier wouldn't fight unless he was told what he was fighting for. And that was almost 200 years ago.

"Gandhi summed up his observations of European culture: 'Christ, yes, but when I see how the Christians behave, then I say no to Christianity.' Dr. Einstein and I were witnesses in 1930 when the youths marching through the streets in Berlin in brown shirts were singing, When Jewish blood spurts from our knives, then things go twice as well.' These were youth who on Sunday stood in churches for the worship of Christ. I read that Cardinal Bertram of 'Breslau had said in a conversation to Pius XII, 'I allow my young people to belong to the Hitler Youth Movement and go to the Nazi rallies, as long as they come on Sunday to the church and receive the Sacrament. I know Protestant ministers who had put *Mein Kampf* beside the Bible on the altar."

"I have no doubt," said the minister, "that religion gets a low mark in Germany during the Nazi time."

"Therefore we should found a cosmic religion," I said, "to form the cosmic man and thus finish, once and for all, the jealous-ies among religions with the 1 am better than thou' concept. I go to the synagogue and to churches of all creeds. I am a Quaker, I am a Swedenborgian, I am a Yogi, and I am a Christian Scientist, which, thanks to Mary Baker Eddy's books, enabled me to experience some wonderful healings."

"With the grace of God," added the minister. "Of course, with the Grace of God. I love Jesus and his mother Mary, too. Dr. Einstein, I, Jesus and Mary can even trace our lineage back to the Royal House of David. When I die I wish to be wrapped in a Fransicscan robe, as well as in a Jewish *tallis*, which Moses, Isaiah and Jesus wore when they prayed."

Einstein chuckled, "O, then you became a Franciscan? You mentioned that desire the last time we met."

"Not formally. I was told that I must first be converted to Catholicism, and I am still bitter about the Church's role in the last war. Nonetheless, my heart wears Francis' robes."

Einstein nodded, "Yes, you could say with Francis, 'Make me an instrument of your peace.'"

"Dr. Einstein, if any religious label could be attached to you — and I feel I am qualified to judge, because I have embraced a dozen Churches to break down the walls for a cosmic religion —it would be 'Franciscan Jew.' When I was received by the Director of the Institute, Dr. Frank Aydelotte, the first time I came to Princeton to visit you, I was told this: You were asked what salary you wanted to have, and you answered, Three thousand dollars a year, or can I live on less?' Come with me to Assisi, Brother Albert, and we'll rent a hut."

We all laughed. I continued, "You spoke once of a fusion of philosophy and science."

"Yes," nodded Einstein, "philosophy is empty if it isn't based on science. Science discovers; philosophy interprets."

Reverend James' unbending zeal gave signs of bursting forth again. "Dr. Einstein," I hastened to say, "you once said that Schopenhauer's statement was among the most profound ever made: 'Man can do what he wills, but cannot want what he wills.' If you believe that, then you believe in determinism. Does this mean that Germany can't be held responsible for its crimes? And Spinoza, whom you venerate, once remarked: 'A stone in the air would think itself free if it could forget the hand that threw it."

Einstein replied, "We are driven by some inner force which we are not always aware of, so I hesitate to talk about free will."

"It appears," I said, "that there is a national mind which is susceptible to political propaganda and nurtured by the subconscious longings of the average citizen. As Hitler perceived, "We Germans think with our blood.""

"Yes," said Einstein, "the Germans had no free will, for they were dominated by their training, environment, and military concepts. However, when they lost a war, these traditions supposedly broke down and a new free will grew up. And how did they use it? They chose Hitler. It is time they learn that they owe to the Jews three things: moral principles, Greek logic, and the refinement of their own language. Without the Bible, there would have been no Luther. I'm afraid Germans have never developed true religious feelings. They bring forth leaders of great power but of no moral responsibility."

"What do you mean by religious feelings?" I asked.

"Love your neighbor for he is like you.' Perhaps it is too simple for their minds, or perhaps it is repugnant to them because Moses said it." He added rather solemnly, "The Germans worry me. They don't understand elementary justice. And the traditional religions worry me. Their long history proves that they have not understood the meaning of the commandment: Thou shalt not kill. If we want to save this world from unimaginable destruction we should concentrate not on the faraway God, but on the heart of the individual. We live now in an international anarchy in which a Third World War with nuclear weapons lies before our door. We must make the individual man aware of his conscience so that he understands what it means that only a few will survive the next war. This man will be the cosmic man." He smiled at the minister, "I believe that is what God wants us to do."

"Dr. Einstein," I interrupted, "since I began teaching at San Jose I have tried to introduce the concept of the cosmic man. The students are remarkable; my classes are overflowing, but what an enmity I have evoked from teachers, especially in my department."

Einstein chuckled, "You probably live in the hornets' nest of mediocrity. I went through that in Germany. There they work with the methods of fear and artificial authority in the schools. In America they work with the method of smiling their problems away. Never mind, continue to rock the boat of complacency. You went through Verdun—what does it matter to you if you have a handful or so of enemies? Your responsibility lies with the students, not the other teachers. A good teacher is not filled with book knowledge, but with a living spirit."

"And this is God," the minister threw in.

"A teacher," Einstein went on, "has to make clear fundamental ends and valuations, and therefore he is a deeply religious man when he accepts this as his highest duty. If we want to save the world we need educators of powerful personalities. Objective knowledge is not all that students need. We need responsible development of youth so that they will understand that service to mankind must be their goal."

I interrupted, "Yes, Dr. Einstein, and not service to the three idols of the American mind: Money, food and fun!"

Einstein looked at me for a moment and said, "I know you are having a tough time there at San Jose, but don't betray your mission. You must help change the world or its future will not be fit to live in. Never make a compromise with mediocrity, for it reacts with greed and brutality to keep the status quo."

As we talked, the minister thumbed through his Bible. "Here it is," he held the book up, "the passage where Joseph forgives his brothers who sold him into slavery by saying, 'for God did send me before you to preserve life.' We don't know why, Dr. Hermanns, you must suffer among the teachers in your department, but God has a plan. I consider you a Joseph, too, Dr. Einstein, although many people think that Hiroshima must weigh heavily on your conscience."

"I have heard these things before, many times," he answered wearily. "Miss Dukas, my secretary, can show you piles of correspondence about this. I happened to have nothing to do with the actual research and development of the bomb. My letter to President Roosevelt was nothing but a letter of introduction for Dr. Szilard who wanted to create adequate contact between scientists and Washington regarding the Manhattan project. I had only handled the problem of nuclear defense when it was reported to me that the Germans were working on such an atomic bomb and, in fact, had uranium mines in Czechoslovakia in their control. I felt it was imperative for the United States to proceed in the development of the bomb, before Hitler used it to destroy London. I also felt that we had to show Germany the power of America, for power is the only language barbarians understand. And when I later learned that the bomb had been created and was to be used against Japan, I did all in my power to avert President Truman from this plan, since publicly dropping it on an empty island would have been sufficient to convince Japan or any nation to sue for peace.

"You from the pulpit," he continued, "may tell your flock that Einstein has read the Bible and follows with his whole heart the words 'God sent me before you to preserve life.'"

"Dr. Einstein, this must have been in your thoughts when you changed from being an absolute pacifist to a realistic pacifist," I offered.

"Yes!" Einstein agreed vehemently. "I would be a traitor to my conscience if I allowed the world to fall into the hands of a dictatorial maniac who was bent to make Europe into one graveyard. I have experienced the German boots, and woe to the country against which they march. I would be a coward if I preached pacifism while youth are taught to commit evil in the name of the state and to march to destroy me, my family and other innocent people."

After a reflective pause, Einstein spoke quietly, "Now we have the bomb and it is my deep conviction that the secret should be confided to world governments, and the United States should immediately announce its readiness to do so. I believe that the Soviet Union, Great Britain and America —now the three strongest military powers — should immediately direct all their efforts to creating a first draft of the world government. That is the only way to interfere in countries where minorities rule over majorities with force and terror. We must change the history of states and nations which is written with the blood of the masses, who had to obey a ruling few. No nation should ever try to get its security by building arms, but by adhering to democratic principles and supporting the creation of a world government, so that at last we will have peace and a new era for the history books."

"Dr. Einstein," I said, "your ideas about the United States working together with the Soviet Union are quite controversial. At the missionary camp where I met Reverend James, the students asked if you were a communist."

"I have given a sufficient answer in my reply to the members of the Russian Academy, which can be read in the Bulletin of the Atomic Scientists of last February. I believe that a planned economy is necessary, but it must not be turned into a fanatic religious belief. Capitalism is not to be seen as the root of all evil, but man himself. We must change the heart of man."

"Speaking of changing the heart of man," I said, "your spiritual coworker, Elsa Brandstroem, has died."

Einstein opened his mouth in astonishment. "The angel of the Prisoners of War?"

"Yes, on the Fourth of March this year. I was especially close to her, since she was the honorary president of the Prisoner of War Association in Berlin. It was she who encouraged me to finish my manuscript on Verdun. She even promised to help me publish it in Sweden."

Edgar blurted out, "For the Nobel Prize?"

"She did not say that," I replied, "but she insinuated that she would write the foreword to it."

"I have heard her name," said Reverend James, "but it was with regard to the Save the Children Fund and CARE packages for Europe after the war." He held up his Bible, "My church collected money and clothes for her charity work."

I motioned for him to put the Bible down, saying, "Elsa did not preach. She told me once that when in Siberia, bringing clothes and food to the starving prisoners, she heard that one of them would be executed for theft. She went to the governor of the Siberian district where the prison camp was to plead for his life. He refused to listen. But when she presented him with a valuable Russian Bible, he said, "All right, you can have him.' Not long after, she received a New Testament from a Jewish urchin, who sold it to her to buy some food."

Einstein chuckled, looking at Reverend James, "Oh yes, I believe that book is about some Jews who lived two thousand years ago."

"That reminds me, Dr. Einstein, that Elsa told me that she had even besieged the first wife of the Kaiser, Empress Auguste Viktoria, to finance her mission in Siberia. The Empress wanted to give her a pearl necklace to support her cause, but then the Kaiser donated a hundred thousand marks from his own pocket."

Throwing a side-glance at Reverend James, I added, "The Empress was the highest representative of the Protestant Church, and yet she was such an Anti-Semite that she never visited the Jewish hospitals, filled with wounded soldiers, during the war."

Einstein smiled, "Yes, the Empress had, nonetheless, a Jewish heart specialist, Professor Israels, whom I happened to meet once. I was probably persona non grata because I pleaded for peace in Holland and Switzerland. The Kaiser and you, Dr. Hermanns, are proof of how German youth are educated to blind submission to the State. The Kaiser's hundred thousand marks could have been better spent on preventing a war."

"One of the closest friends of the Kaiser," I said, "was Albert Ballin. He had warned him not to listen to his generals, but he was a Jew and hated by the Empress."

I took some papers out of my pocket. "At a Christmas dinner in her Cambridge home in 1945, Elsa gave me this copy of her speech to the World Church Conference for Practical Christianity of 1925, which she called 'Charity as the Reconciliation for the Peoples of the World.' Allow me to read one paragraph, which reminds me of your great respect for intuition:"

With the ethical development of individuals and nations, is also the opportunity of doing noble and sincere works of charity. But this development does not go hand in hand with the rational, but with its opposite. If we would try somewhat less to think and a bit more to feel, we would certainly go farther.

This speech made a great impression in Geneva. Not only Stresemann commented on it in my presence when Germany entered the League of Nations, but also Fridtjof Nansen, with whom I had tea. Nansen had continued Elsa's work in Russia and received the Nobel Peace Prize.

"She told me," I continued, "that she hoped to have seen Einstein in Cambridge when Churchill was awarded his honorary doctorate at Harvard. 'When you meet Einstein,' she said, 'tell him that I, too, believe in one world.'"

As I handed her speech to Einstein, he said, "Indeed, it is not intellect, but intuition which advances humanity. Intuition tells man his purpose in this life."

Reverend James asked, "And what about the next life?"

"I do not need any promise of eternity to be happy," replied Einstein. "My eternity is now. I have only one interest: to fulfill my purpose here where I am. This purpose is not given me by my parents or my surroundings. It is induced by some unknown factors. These factors make me a part of eternity. In this sense I am a mystic, and Elsa Brandstroem appears to be inspired by those unknown factors which mold our inner self." Looking at me, he added, "Your Verdun book should make the world think twice before plunging into another Verdun, but the next Verdun will be global."

The minister's reference to the biblical passage of Joseph earlier and now Einstein's mention of purpose had reminded me of Jung's concept of synchronicity tied to teleology, his belief that within each experience is the seed of a finer purpose. I realized that Einstein himself demonstrated such a theory. I reminded him of the compass he had been given as a child, and suggested that it was proof of the synchronicity of events. "Without the compass," I said, "you would not have started to make an inquiry about causal processes in creation, and without being ill as a small boy and lying in bed, the compass would probably never have been given to you. And without your uncle being interested in physics, you might not have become The Einstein. But then, your mother had predicted that you would be famous."

Einstein looked bored, yet he chuckled, "What else to put me in the limelight?"

"No," I responded, "I put Jung and his theory in the limelight." Einstein looked at me and said, "Jung is a great man; too bad that he accepted the presidency of the Psychoanalytical Institute in Germany, though the father of psychoanalysis, Sigmund Freud, and other Jews in the field were forced to resign or flee under the Hitler terror. In his behavior he followed the example of most German scientists, though as a Swiss citizen he could have safely protested and formed a new institute in Switzerland. But I agree with him that causal methods in physics are not enough to explain the laws of the universe. My concept about relativity had to do with my feeling rather than my intellect."

"You affirm" I said, "what Jung called 'synchronicity' of events, whose causality cannot be explained by intellect. I learned that intuition comes from the depth of the psyche. That I should come to tea when that man stood in front of your door and threatened you, was also synchronicity of events."

Einstein interrupted vehemently, "Don't talk to me about that man. I wrote you my opinion about it, didn't I?"

"Yes," I said, "but in the whole world you may be the only one who thinks the man was harmless. But now let us examine my own life. Can I not say, too, 'God did send me before you to preserve life.'? I volunteered to fight the Kaiser's war, and soon I was to experience what Jung calls the synchronicity of events. The eyes of two children who stood before their burning barn in an Argonne village followed me, and I stopped singing: Victoriously, we will crush France'—so, too, I stopped believing what I learned to sing in the Officer Training School:

'We will hate because we must hate.
We will hate because we know how to hate.

We laugh together;
We hate together.
We hate together our arch-foe, England!'"

"But there is a judgment after this life!" cried the minister.

I felt a chill as if ice cubes were dropped down my neck.

Einstein went on: "Let it come for those who believe it. I cannot conceive of anything after my physical death—perhaps it will end it all. The knowledge that I am now on this earth and a mysterious part of eternity is enough for me. My death will be an easy one, too, for since early youth I have always detached myself from family, friends, and surroundings. And should I live on, I have no fear of the next life. Whatever good I did helped to free me from myself. What a miserable creature man would be if he were good not for the sake of being good, but because religion told him that he would get a reward after this life, and that if he weren't good he'd be punished."

"But isn't that also true of the Jewish religion?" I asked. "Think of the Ten Commandments and of how they are enforced by fear."

"No," said Einstein. "The Commandments were given to make the Jews a holy people and isolate them from the surrounding people, who believed in many gods, human sacrifice, and immoral practices. The Jew's religion had one God, and for this they are hated to this day. Hitler announced that the greatest duty of the Germans was to overthrow the belief in one God and the Ten Commandments because they were Jewish concepts. Hitler wanted to become the German god. The value of Judaism is in its spiritual and ethical content and in the corresponding qualities of individual Jews. In spite of their persecution by the Christians through two thousand years, the Jews have never given up their love for the spiritual cultivation of their thinking."

The minister promptly added, "But Paul has told the Philippians, "Work out your salvation with fear and trembling.'"

Einstein turned to me and said in German, "I don't know why men always want to convert me to their beliefs."

"Your name is good advertisement."

"If I were allowed to give advice to the churches," Einstein continued, "I would tell them to begin with a conversion among themselves, and to stop playing power politics. Consider what mass misery they have produced in Spain, South America and Russia."

Einstein returned to English. "I don't deny for a moment that the churches have done much good in a humanitarian sense, but I'm sorry to say that their spirituality has often been perverted. Much too frequently, in order to preserve and augment her own power, the church has abased herself before political regimes."

The minister interjected, "There has always been ample protest in my church against politics in religion."

"I have written a poem," I said, "in my anguish over the Church's complicity:

CHURCH

O Church, your tale, how sad and grim,
You washed your hands in Pilate's water.
The Children's Crusade you have blessed.
Your yellow star inspired him,
the swastika chief, to slaughter
the sub-humans. His hate-filled breast

could boast, "I learned my lesson well
from Popes." O Church, O crimson spot!
No cross, put on your highest tower,

can change the truth of heaven and hell:
The power of love is blessed by God,
but not the devil's love of power.

Einstein looked at me, "That is why the most beautiful Church for me is the church of conscience, found in the silence of one's own presence." And when he saw our wondering faces, he added, "Unselfishness, humaneness, service to your brother—these are the values which the Church should practice for once, instead of constantly trying to gather in more souls. A cosmic religion is the only solution—then there will be no more Church politics of supporting the mighty at the cost of the human rights of the poor."

"I made up my mind on the way to the trenches never to aim against a Frenchman, but rather to be killed first. I then preferred to be a traitor to the nation than to my conscience."

"Bravo," mumbled Einstein.

"I'm not so arrogant as to say that God used me," I continued. "But this I know: I was used by a power outside of myself. At Verdun, when all were dead around me, and I was half buried by a shell, I cried, 'God save me, and I will serve you for as long as I live!' I was saved, and in the next couple of days I was responsible for saving the lives of several hundred German soldiers."

There was a silence; then Einstein looked at the minister and said, "My God may not be your idea of God, but one thing I know of my God—he makes me a humanitarian. I am a proud Jew because we gave the world the Bible and the story of Joseph. As long as I live, I will try to save lives."

"But your efforts may be limited by the ever-threatening atom bomb," the minister replied.

"Do you think," I asked, "that the next bomb will finish us all?"

"I don't think so. Two-thirds or more of the world's population would be killed, but enough would be left to start again."

My nephew Herbert asked whether there was any way to prevent such destruction.

"Yes," said Einstein, "if we can only subdue man's evil spirit. We scientists won't change the hearts of other men by mechanisms, but by changing our own hearts and speaking bravely. We must be generous in sharing knowledge of the forces of nature, but only after establishing safeguards against abuse. We must realize we can't simultaneously plan for war and peace. When we are clear in heart and mind, only then shall we find the courage to surmount this fear which haunts the world."

"Do you believe," asked Edgar, "that we can keep the bomb to ourselves?"

"Certainly not," said Einstein. "We must attempt to create a supranational government so that no secrets are necessary."

"Weren't you the chairman of the Emergency Committee of Atomic Scientists?" I asked. When he verified this, I went on, "Weren't they developing the atomic bomb?"

"Some of the members have worked on the Manhattan Project, but they are now concerned for the continuance of life on this planet. The committee was established to disseminate the simple facts of atomic energy and its impact on society. The atomic question is not so much a scientific as a moral one, you know. It disturbs me that, even though the atomic age is here, people still don't want to change their mode of thinking.

"Also, when I look around I see a new slavery for the individual developing in the United States. Everything seems to be in preparation for war rather than peace, glorifying the warlike spirit to cope with the Russian threat. This capitalistic interest in armament production reminds me of the conspiracy of the Krupps and others with Hitler. War industry is a source of wealth. These industrialists owned castles, land, and yachts."

I quickly said, "Henry Ford had printed and published throughout America that anti-Semitic pamphlet, Protocols of the Wise Men of Zion. He took over the Dearborn Independent newspaper to propagate more of his anti-Semitic ideas. Bruening had told me that Ford made enormous donations to the Nazis through secret channels arranged through Richard Wagner's son Siegfried and his wife Winifred, who lent respect and stature to the Nazi cause."

Einstein nodded his head, "I remember that Max Liebermann told you about Wagner's real father being a Jew. Did his children know? Of course, like good Germans, they didn't want to know. And as for Ford, one sees the danger industrialists present when they have tasted success and become nationalists. 'Money talks,' as they say here. True, America is a democracy and has no Hitler, but I am afraid for her future; there are hard times ahead for the American people, troubles will be coming from within and without. America cannot smile away their Negro problem nor Hiroshima and Nagasaki. There are cosmic laws."

"Mr. Einstein," said Reverend James, eager to change the subject, "atomic power is mysterious, and God is mysterious, too."

"What are you telling me when you say this?" Einstein turned to face the minister. There is no true science which does not emanate from the mysterious. Every thinking person must be filled with wonder and awe just by looking up at the stars."

"But wonder and awe are not enough," retorted the minister. His ascetically handsome face showed the fervor of an evangelist.

"I know," said Einstein, looking at me in the same quizzical way, which by now had grown into a tacit reproach. "Do I have to repeat that I don't believe in a personal God who rewards and punishes His creatures? He did not create cosmic laws in order to override them when man asked Him to do so."

I asked whether it wasn't the mission of the Jews as a chosen people to atone for the sins of the world.

Einstein replied, "I can't comprehend such mysticism, but if you feel gratified by such speculations, go ahead." He shook his head. "When I think of the tragic behavior of the German intellectual elite — famous scholars, ministers, and priests, generals — "

"Yes," I interrupted, "and many of them were churchgoers, like Brauchitsch, Rundstedt and Rommel."

"When I think of this," he went on, "I can only say that the welfare of humanity must take precedence over loyalty in one's own country or to one's church. I repeat, we need a cosmic religion."

"What do you mean by cosmic religion?" the minister asked.

Einstein leaned forward, "Dear Reverend, it is not a religion that teaches that man is made in the image of God —that is antropomorphia. Man has infinite dimensions and finds God in his conscience. This religion has no dogma other than teaching man that the universe is rational and that his highest destiny is to ponder it and co-create with its laws. There are only two limiting factors: first, that what seems impenetrable to us is as important as what is cut and dried; and, second, that our faculties are dull and can only comprehend wisdom and serene beauty in crude forms, but the heart of man through intuition leads us to greater understanding of ourselves and the universe.

"My religion is based on Moses: Love God and love your neighbor as yourself. And for me God is the First Cause. David and the prophets knew that there could be no love without justice or justice without love. I don't need any other religious trappings."

The minister broke in, "Oh, the depth and richness of God's wisdom and knowledge."

Einstein smiled. "When man acknowledges that he can know all, but not just yet; when he has the humility to think himself not so important, and considers himself only a grain on the shore of infinite wisdom, then he is religious. In this sense, I belong to the ranks of devoutly religious men."

"Amen," whispered Reverend James. "'Without me, says the Lord, you can do nothing.' You know, Dr. Einstein, your humility comes from the spirit."

"I believe the main task of the spirit is to free man from his ego," said Einstein.

"Dr. Einstein, ridding oneself of one's sinful ego cannot be done without the help of the spirit." He opened his Bible. "For as Paul wrote to the Thessalonians: We know, brethren, beloved of God, that He has chosen you; for our gospel came to you not only in word, but also in power and in the Holy Spirit and with full conviction."'

Einstein looked helpless and harassed. I waved my notepad at the minister, trying to silence him, but he only returned a smile, as if to say, "Let me go on, for this is my hour." I rose from my chair, ready to give the missionary angel some wings to fly away with, and then felt impelled to look at Einstein again. Had he sensed my thoughts? He smiled. I sat down.

The minister continued, "Dr. Einstein, you believe in great examples, and Saint Paul is mine. He said to preach the word 'in and out of season.'"

"But I'm afraid you are employing your zeal on an unsuitable object."

"Dr. Einstein, you have a compassionate heart and will agree that there is nothing but suffering around us. But that can be made meaningful and bearable when we accept it as Christ did on the cross."

"The cross was used to make wars, to make warriors, to make a concordat with Hitler. I feel that such a cross has lost its meaning."

There was a silence, and I took advantage of it. "Dr. Einstein, the students at the missionary camp were shocked by your support of Bertrand Russell and his attacks on the Christian concept of marriage."

"I don't care for Russell's personal views," he answered abruptly, "but that this great mathematician and philosopher should be dismissed from his chair in New York smacked of Hitlerism."

"They resented," I continued, "the fact that you, a refugee, sided with that 'much-married nudist,' as they call him." I motioned to my nephew that we should go. While we were shaking hands, Herbert pointed to his camera, and I asked Einstein if he would mind having his photograph taken.

"All right," he said, and promptly stood in position behind his desk. But the room was too dark, and we moved outside.

Hilda and her daughter Margot joined us as we were taking pictures (see Fig. 3). When I presented them, Einstein said, "One can see you come from Germany. There is something in you that is not American—your attitude, your dress, the way you carry yourself." With a broad grin he added, "But then, after all, we are all alike, for we are all derived from the monkey."

As Einstein was ascending the stairs to hide himself in his office, I rushed after him and caught up with him just inside the corridor. "Professor, I owe you an apology."

He turned around. "Yet another one?"

"Unfortunately, yes. I am not responsible for the minister's zeal."

"Don't worry about it. He is sincere and Fm used to it. There will always be people who want to help me into heaven. Even a Cardinal and a Rabbi have been concerned about my soul. Your minister plucks God from man's soul, and I from nature. To each his own."

"Still, I feel like a wicked child."

He smiled and shook my hand. "Better a wicked little Willi than a wicked man."

I pulled from my pocket my essay. "Here is 'Churchill at Harvard.' You asked me to write it."

Einstein looked helplessly first at me and then at the essay.

"It's five years old," I said, "but I wanted to give it to you personally, lest it land m the wastebasket with the majority of your daily mail."

He took it and to my amazement read the last page and scanned through the others, from the back forward. "Did you give Churchill your poem *Dunkirk*, too?" he asked.

"I was told that President Conant of Harvard brought the essay along with the poem to Churchill."

Einstein smiled, "I make rhymes, too, but you're better."

He stood before me so unassuming, so ageless, so from another world, that I asked, "Dr. Einstein, do you believe in reincarnation?"

He swiped the air with my essay as if to slap me. Did he know, or rather feel, what I was going to say, that he reminded me of a Jewish prophet?

"I thought you wanted to tell me something interesting," he said, now switching to German.

Photograph of Einstein and the author. Third Conversation, September 11, 1943.

I spoke also in German, "O yes. Dr. Einstein, some years ago I had a luncheon with the Queen's brother Sir Bowes-Lyon in Washington at the English Embassy."

"Ah," grinned Einstein, "you gave him your Dunkirk, too."

"Yes, I did and three other poems: My anthem *Free Men Go Forth, England,* and *The Women of England.* But I'll not speak about myself."

"I see," he chuckled.

"Dr. Einstein, am I so amusing? Or do you feel what I'm about to say?"

"Speak. Lighten your heart."

"Sir Bowes-Lyon regretted that you declined the offers to stay in England. You would have been ennobled."

"O," he joined in, "Lord Einstein."

"Why not?" I said. "Your friend Sir Herbert became Lord Samuel."

"Tell the Queen's brother, should Einstein come back to earth, he wouldn't mind—no, tell him that he will become a shoemaker." He grabbed my hands and pressed them. "Dr. Hermanns, if you want to do something for me," he said with a faraway look, "tell the world that if I had foreseen Hiroshima and Nagasaki, I would have torn up my formula in 1905. Help me to create two things: a World Youth Parliament and a Supranational Government with the only allowable military force. The international peace-keeping force that you talked to me about during the war is now timely. You are a sociologist and still young. Make it your holy duty. You were at Verdun. Make your life meaningful—poems are not enough. We don't live anymore in the time of Goethe, Schiller and Heine. Give your life a worthy purpose —not just a goal but a worthy purpose."

I was overwhelmed and said, "But you may help the world with a new formula."

"Whether this grace will be granted to me, I don't know." He looked forlorn and rubbed his hand through his hair.

"Dr. Einstein, we stand here in Princeton as we did in Berlin in

1930."

"Yes," he mocked, "but this time no pulling me to the police."

"No," I said. "We are in America."

"America, however," mumbled Einstein, "uses Russia now as a pretext to arm and create more terrible nuclear bombs. If I were young, I would leave the United States. I want to live where scholarship is free and unattached to the military machine. I want to live where spiritual values are not suppressed by the State. Nothing has real value which is not done out of love for one's fellowman. Poor America—the Apocalyptic Rider is coming."

As in 1930 before his door in Berlin, he disappeared without saying goodbye.

The next day at tea the Reverend James began, "When we visit Einstein again, let's stress historical Christianity: Sinai, the Road to Damascus, the open tomb..."

I quickly discouraged him. "We'd better resign ourselves to the fact that God has many mansions. And one I'm sure is for Einstein, so he can go on studying the universe. You know, Reverend James, he said that you were sincere."

Reverend James smiled sadly, and I knew my words were poor consolation for a young missionary, especially when a hundred others in that Canadian camp had envied him the opportunity of bringing Einstein's soul back on a spiritual platter. Remembering Einstein's last words to me about the fate of the United States, I shared them with the minister. In a moment's time, Reverend James pulled out his Bible and read the passage from Revelations:

And I looked, and behold a pale horse:
and his name that sat on him was Death,
and Hell followed with him. And power
was given unto them over the fourth
part of the earth, to kill with sword,
and with hunger, and with death, and
with the beasts of the earth.

Reverend James' fire changed to a soul-warming love when he went over my notes with me, patiently helping me complete statements of our conversation that I had scribbled speedily down the day before. Thus with the minister's assistance I was able to share my latest experience of Einstein with my students in San Jose, whose fraternities and clubs opened their doors to me.

Fourth Conversation With Einstein

1954

Since I was embracing all religious movements and spiritual paths, I also had Monsignor, later Archbishop, Fulton Sheen as a friend. He knew of my intention to found a cosmic religion including the Jewish, Christian, Vedic, Buddhist, and Islamic traditions. He wanted to devote a broadcast to anti-Semitism and needed permission from Einstein to give a special lecture on him and his work, since Sheen felt that he had to make up for his and other Catholic leaders' derogatory remarks about Einstein.

In the summer of 1954 when I again was visiting my sister Hilda in New Rochelle, my friend William Miller, an editor on the staff of Life Magazine in New York, offered to drive me to Princeton. He brought along his eldest son, Patrick, a physic's major at Harvard who was unsure of his purpose in life, was frustrated, and had lost interest in his studies. Bill hoped that Einstein might help revive him.

My zeal was dampened when Einstein's secretary told me over the phone that the professor had a cold and would not receive visitors that weekend. Since secretaries often play Cerberus, however, I withheld her *non possumus* from my friends. But then the dark clouds increased: Bill had car trouble and was two hours late calling for me, and once on our way, we missed the exit on the New Jersey Turnpike. All this meant that we wouldn't arrive before noon as planned, but at a time when Einstein, true to European custom, would be having guests to tea. But we persevered, and clung to Bill's intuitive advice: "Think positively, and well produce such strong vibrations that he can't refuse to see us."

We reached Einstein's home late Saturday afternoon. I jumped out of the car and rang the bell, and a plainly dressed middle-aged woman opened the inner door. She informed me that Einstein had guests for tea, and that I should return the next week after making an appointment. I recognized that this was Miss Helen Dukas, his faithful secretary, and that she was immune to our positive vibrations.

The screen door through which she addressed me seemed a barrier to the sanctuary within. She wasn't impressed by my speaking German, her native tongue, nor by Pat's predicament, nor by my need for a statement from Einstein on anti-Semitism. But when I told her that I knew him personally and had once saved him in Berlin from the threats of a crazy writer by taking him to the police for protection, her resistance diminished. She finally opened that formidable door and ushered me in.

Behind a curtain that separated us from the next room I could hear the clinking of teacups and Einstein in conversation with several people. Would he remember my name and come out? He was seventy-five years old and daily plagued by visitors and letters from all over the world. Furthermore, his secretary, tiny and slim though she was, seemed to assume such proportions that she filled the whole room, leaving no space for my friends and certainly none for Einstein. She didn't budge; she wanted to know the questions I intended to ask. The curtain became yet another barrier, another screen door. Helen Dukas, in one past life, must have been Archangel Michael, guarding the gates of Heaven with her flaming sword.

I happened to have with me a religious essay of mine, Mary and the Mocker, which had just been published by Our Sunday Visitor Press (Huntington, Indiana). I handed it to her, and she retired. The pamphlet seemed to animate conversation behind the curtain, and I heard someone say, "Did you notice that Fulton Sheen wrote the foreword?"

While I waited, my eye fell on a peculiar altar-like arrangement, on which stood a statuette of a mother with a child in her arms. For a moment I was startled —this was a meaningful coincidence, which like Ariadne's thread has always shown me the way out of the labyrinth of my life: Mary and the Mocker in the hands of Einstein in the next room, and the statue of the Madonna in this room. The Madonna was Jewish, yet this was still the home of

Einstein the realist. But then I remembered having seen many such figurines at his apartment in Berlin a quarter century ago. This exquisite altar stood out amid a few worn wicker armchairs, a simple round table, and a narrow cot that looked most uncomfortable, as though a monk did penance here.

The curtain was pushed aside and Einstein appeared. He smiled and extended his hand. "It seems that we have met before." I apologized for having taken him away from his guests, and he assured me that he had finished his tea and asked what I wished to know,

Again, a delicate situation. If my question weren't interesting, he would answer it briefly and then bow out; but if Bill's son had the chance to ask his question, and if it happened to interest Einstein, then time wouldn't exist for him. But seeing Miss Dukas' dark eyes somewhat reproachfully fixed on me, I became so nervous that I wasn't able to look at my agenda.

"Dr. Einstein," I blurted out, "stupid as I may be in Physics, I have still something in common with you: The love for the Madonna."

Miss Dukas, standing by the door, snapped, "But we don't worship her!"

"Dr. Einstein, you are a mystic, and so am I."

"Wait a minute!" Einstein waved his hand. "I am not a mystic. Trying to find out the laws of nature has nothing to do with mysticism, though in the face of creation I feel very humble. It is as if a spirit is manifest infinitely superior to man's spirit. Through my pursuit in science I have known cosmic religious feelings. But I don't care to be called a mystic."

"You are a child, then," I replied. "A child is not a mystic; it simply looks around to understand its surroundings. Our environment, however, includes the stars. We are part and parcel of other worlds. I discovered at Verdun that there are more dimensions to man than the three dimensions of space. Man has infinite dimensions. That is why I am interested in mysticism, or metaphysics, as a way to understand better what my existence is all about. You once told me that moral obligations are the most important of all human problems. This Jewish Madonna is not here by chance. You attracted it as a symbol of love." I turned to Miss Dukas. "Dr. Einstein also taught me that if we can penetrate a little deeper into the eternal mystery of nature, we will be granted great peace. When I lost my mother as a boy of seven years, a shoemaker's widow tried to comfort me by taking me to her little bedroom, and then pointing at such a statue said, 'Willi, she will be your mother.' Fourteen years later when I was in French captivity, some Catholic German prisoners wanted me to translate a French newspaper article about the apparitions of Maria at Fatima. I refused, saying it was only religious propaganda and superstition. Many years later, in fact just after my last conversation with you, Dr. Einstein, I was invited to see a statue of the Fatima Madonna, and as I stood before it she seemed to speak to me, 'Go home and amend.™

Einstein, handing me back my booklet, said, "And this story you called Mary and the Mocker?"

"The story is just the introduction to a long poem I wrote about the apparition and its effect on a communist who only believed in scientific materialism."

"Fulton Sheen is known to be a dangerous converter. Did you become a Catholic?"

From the edge of the couch, Miss Dukas asked, "Aren't you the gentleman who brought a Protestant minister here to convert Dr. Einstein six years ago?"

"I didn't give this book to convert you, Dr. Einstein. I am a Jew to the Jews, a Catholic to the Catholics; I am all things for the sake of love and understanding. It was you who inspired me to become a member of your cosmic religion. And more: In our first conversation in 1930..."

"Yes, yes," interrupted Einstein with a smile. "Don't remind me of that."

I continued, 'You asked me then how to change the heart of man. I thought about it for over twenty years and want to discuss this with you."

"First you tell me if you've finally become a Franciscan."

I now smiled, "Yes; after Sheen baptized me I had the proper prerequisite, so a couple of years ago I was professed as a Third Order Franciscan in Assisi at the tomb of St. Francis."

"I then congratulate you," nodded Einstein. "I hope you can influence the Church."

"I hope so, Dr. Einstein. Didn't you state that the Church was the only opponent of Communism?"

"I don't believe I have to emphasize that the Church at last became a strong opponent of National Socialism, as well."

His secretary added, "Dr. Einstein didn't mean only the Catholic Church, but all churches."

I was amazed at Miss Dukas' impromptu remark, but no doubt this alert woman wanted to nip in the bud any tendency to play up the Catholic Church. She went on, "You have some people waiting outside."

"Why don't you show them in," said Einstein.

"First allow me five minutes, Dr. Einstein," I spoke German now. Smiling at Miss Dukas, I added, "You, too, may be interested in what happened to some people we both have known."

Einstein, offering me a chair, sat down. I said, "You said once that changing anyone is a superhuman task unless we use arguments which appeal to the emotions as well as to the intellect. How true! I had a long conversation with Cardinal Innitzer of Vienna a few years ago. He tried to rationalize his pro-Hitler stand, saying that he had no precedent from the Vatican to act differently."

"Yes," nodded Einstein, "he acted according to the spirit of the Concordat."

I added, "Yes, but when I seized his emotions by relating the loss of my family members in the gas chambers, he almost wept and said that he also felt guilty —adding that the whole Church, beginning with the Vatican, should feel guilty."

"Emotional responses don't lie; intellectual ones often do," said Einstein. "But a combination of the two are like two eyes to view the clear picture."

Miss Dukas had sat down on the edge of the couch, and this caused me to think that for refugees, no stronger bond exists than that of their mother tongue. Her role of the archangel had given way to the interest of a German in her homeland. Looking at both of them, I said, "I have some first-hand information from my last trip to Germany about some of our acquaintances."

Einstein looked into my eyes, but he gave no answer. I went on, "Did you know that Bruening returned to Germany in 1951 to teach political science in the University of Cologne?" Einstein nodded his head. "I was told he was so scandalized about the general regard of him as a traitor to Germany that he returned to the United States to peacefully live out his life."

"Yes," said Einstein, "he probably fears, as I do, that it is a lost cause —the Germans are un-teachable."

I continued, "General von Schleicher, who plotted against Bruening and succeeded Papen as Chancellor, was murdered in the Roehm Purge of 1934, whereas Papen escaped death by a hair. He had to appear before the Nurcnberg court, where he was acquitted, only to be later sentenced by a German court to eight years in jail. He was dishonored and humiliated. His sentence was rescinded, however, in 1949.

"One man I knew well from the Humboldt Club, Hitler's trusted companion, Albert Speer, received a twenty year sentence from the Nurnberg court.

"General Groener, a man dedicated to preserving the Weimar Republic and who told me how much he despised certain irresponsible leaders in the army, escaped the concentration camp by mercifully dying beforehand. A friend of yours and mine, Max Liebermann, also avoided the Nazi noose by dying peacefully at home before the Gestapo's plan for him could be carried out.

"It is of interest here to mention that Reverend Martin Niemoeller told me at the Twentieth of July Memorial Service in Berlin that one should suspect those Germans who always spoke during the Nazi time of protecting the purity of Aryan blood, for probably they all had a Jewish ancestor in their family tree which they were ashamed of. It was whispered that Professor Lenard, the

Nobel Prize physicist who ridiculed your relativity theory as Jewish science, was one who had a so-called 'Aryan infection.'"

"Niemoeller is a respected theologian," said Einstein, "one who is also dedicated to social justice."

I added, "He spent some years in the concentration camp, as did my friend Probst Heinrich Grueber. To the glory of the Protestant and Catholic Church there were many who were martyred because they had protested against the attacks on Jews and their subsequent removal to concentration camps. Unfortunately, these were still only a handful of Christians, compared to the number holding Church membership who admired Hitler, Mussolini and, let me add, Franco."

"Of all my acquaintances, Gerhart Hauptmann fell from the highest pinnacle to the most ignominious death. Fate chose him to witness the bombing of his beloved Dresden. At the end of March 1945, a month after the leveling of Dresden, he sent a radio address to the German people, in which he expressed the plea, If only God would love humanity more! If only He would purify and give it more grace for its salvation than He has until now."

"How convenient," said Einstein, "to push the guilt onto God. How sad."

"On his deathbed," I continued, "he became suddenly aware that his house in Silesia was occupied by Russians and Poles. His last words that he cried out were: 'Am I still in my own home?'"

"Oh, Dr. Einstein, in 1913 Hauptmann, my idol, wrote a special play commemorating the one-hundredth anniversary of the war of liberation from Napoleon, which the Crown Prince had forbidden because it spoke of one humanity and one love, which sounded anti patriotic to him. What a fall, what a deep fall. I was greatly affected. He could have come, as we did, to America. And he would have been honored —he, as the author of the socially aware play *The Weavers*.

"As for Goebbels, you know, he had his wife give poison to their six children, and then had an S.S. orderly shoot them both. What a change from a Jesuit student to a student of Hitler! He had tried to become a member of the Stefen George Circle, where he was told that he was a journalist at best, and certainly not a poet. This offended him and is said to account for his anti-Semitism. I was told that his one surviving son from a previous marriage, Harold, became a priest to atone for his father's crimes."

Einstein broke in, "I hope the Sermon on the Mount means more to him than it did to his father."

He rose from his chair and went to stand before the old Madonna statue which stood by the window in his room. Then he turned around and looked at me, "Doesn't this prove what I have always said? There is no security and peace unless we have a supranational government. What a challenge for you! Form a world youth movement, as I told you before, with a vision of the cosmic man. We must liberate ourselves from what we have inherited of antisocial and destructive instincts."

"I have brought a program with me, Dr. Einstein. I've envisioned the Anne Frank Academy to be a seed for the World Parliament of Youth. Mrs. Roosevelt has even agreed to act as patron."

I handed him the brochure, and he looked over the list of purposes. That's fine, Dr. Hermanns," now the organizational Battle begins."

"Yes," I replied, "collecting funds is a problem. At least I've managed to have much of what I've written incorporated into the bylaws of the Anne Frank House in Amsterdam."

"I am pleased," said Einstein, tapping the brochure with his finger, "that Mrs. Roosevelt has offered to help you. Her introduction to Anne Frank's Diary has greatly assisted the publication of this young Jewish girl's cry for love in a hate-torn world."

"Do you know, Dr. Einstein, what Mrs. Roosevelt said about you when I visited her in New York? You should get a seat in the United Nations and help create a Magna Carta of Mankind."

Einstein smiled, "You manage to meet some great women, don't you? As I remember, you met Jane Addams of Hull House."

"What a memory you have," I threw in.

"And your dear Elsa Brandstroem and that simple Sister Nanny."

"How good that you speak of these cosmic women. Elsa Brandstroem made a speech in 1925 in Stockholm on the initiative of Archbishop Nathan Soederblom, whom I met in Geneva in 1926 at a garden party given by Jane Addams, to celebrate Germany's entrance into the League of Nations. He told us that Elsa considered love, unconditional love, as the only reconciling power to bring all nations into a world union. He said that Elsa Brandstroem would go down in history as one of the world's greatest women. Unforgettable for me was one of her long conversations with me in Cambridge, when she invited me to live in the refugee home she had founded. She told me of a couple from Vienna, whom she had sheltered there, too. Mr. Schwartz was an organist in St. Stephen's Cathedral. Because his wife was a Jewess, two Gestapos knocked at his door a few days after Hitler's triumphant entrance into Vienna. They came before dawn and roused the couple from bed. The couple dressed themselves and were taken to a jail, where other Jews, men, women and children, had been herded together. Suddenly, Mrs. Schwartz remembered that her little poodle was left behind. When the soup was brought in, she told the prison guard of her concern. Soon the Gestapo officer entered the cell, yelling at her, You played a beastly trick on me, Jew.' He slammed the door and left. An hour later, he returned and said, There is no dog!' 'Oh yes,' she said, 'he was hiding when you came in.' 'Come along, then.' At the house, the officer remained outside as Mrs. Schwartz went to her bedroom. The dog wiggled out from beneath the bed and jumped into her arms for joy.' After their return to the prison, he took it away from her. Caressing the dog, he said, '111 take care of him. He is now my dog. Look how hungry he is because of you, subhuman Jew.' A day later, he brought the dog into the cell: 'The dog does not eat. You must give him the food.' She then gave the dog soup and meat brought by the guards. This continued for several days until they were transferred to a concentration camp. There were two or three children in their prison cell, who hungry as they were, tried to grab some meat from the dog's portion, but they were pushed back by the guard.

Elsa Brandstroem ended the story with the remarkable sentence, 'Willi Hermanns, if you want to honor me, continue the work of love, which preserved you at Verdun.'"

Einstein said, "That is one of the reasons why I support the Jewish state in Israel. The Jews gave the world the Bible. Though the Europeans have told the Jews they should assimilate, the behavior of the Nazis and the Allies, too, has demonstrated that they do not want the Jews."

"The Jews gave the world the commandment, Thou shall not kill,'" I remarked. "The Christians believe, however, that the New Testament cancels the old covenant, as if God would go back on his word."

"Israel will not last," said Einstein, "unless we create a United States of the World. I am afraid that humanity has not learned its lesson and that we will have a Third World War, destroying three-quarters of mankind."

I pointed at the Madonna statue and said, "She was a Jewess. I think she smiles now."

"Yes, Dr. Hermanns, you are a dreamer."

"Well," I said, "aren't all great things dreamt of first? You dreamt, you told me in our first conversation, of catching a sunbeam to get to know it better —the beginning of your relativity theory."

Einstein smiled, "Yes, that's true. And this as a Jew! How differently I would have been treated if I had been a Christian!"

"You know, Dr. Einstein, your colleague, Eritz Habcr, suffered a terrible fate. At the end of 1933, I was on my way to the French Consulate to look for a hiding place in case I didn't get my visa. There was shouting in the street. I saw posters carried by Nazi youth: Tteath to Einstein!' 'Death to Fritz Haber!' 'Death to Bruening!' In those days I wore the ribbon of my Iron Cross in my lapel as protection. Thus, I had the courage to ask two students with their university caps on and who applauded the posters, *Why Fritz Haber? Didn't he do a great service in the World War? Wasn't he awarded the Nobel Prize in Chemistry?' One of the students told me Haber deserved to be killed because his discovery of extracting ammonia from the air had prolonged the German war effort and cost perhaps a million more lives. I responded, 'But what about Einstein then, on whose head you have put some 20,000 DM?"

Einstein smiled. "No. They increased it to 30,000 DM."

"I said to the student, 'He signed a manifesto protesting Germany's aggression in the World War.' 'A stab in the back!' was his answer, 'And both are Jews!'"

Again Einstein smiled, "*Einerseits oder andrerseits*, the Germans can easily take two sides of any issue to prove their point. It's amazing how unaware they are of this; if only they could hear themselves speaking. The idea of 'fair play' cannot even be expressed with the German vocabulary."

"And do you know, Dr. Einstein, what a miserable death Haber died? Though he honorably served his Fatherland as a major in the German army in the World War, he had to flee for his life. A Swiss student at Harvard told me that he died a lonesome death in Basel."

"Yes," said Einstein, "I know. He wanted to emigrate to Israel, but didn't make it."

"Let me tell you about Helmut von Gerlach..."

Einstein interrupted, "Yes. I remember. From the League for Human Rights. Did they get him?"

"No. I met him by chance; yet I should not say by chance. I should say by a coordination of events in Paris. I had just visited the Arc de Triomphe and paid my homage to the Unknown Soldier of the First World War when I bumped into him. He said to me, 'You and I will soon die as unknown soldiers. "Fugitives and vagabonds," as the Bible would say.' He died shortly thereafter in Paris.

"And there is another acquaintance of yours and mine, Rudolf Breitscheid, the leader of the Social Democrats. He fled to Paris and despite my warnings he stayed, only to be hauled away by the Nazis and put in a concentration camp to perish."

That reminds me," said Einstein, "of how I was followed by secret agents, even in a peaceful Belgium spa. The King of Belgium had me watched by his guard as long as I was there."

"Let me add a last name: the Empress Hermina. You will remember, she told me we should elect Hitler; later, when he had 'cleaned out the Augean stable of the Communists, then we come.' She came to her senses too late and suffered a fate similar to that of Gerhart Hauptmann; perhaps even worse. She was hauled away from her castle in the Harz Mountains and, I was told, was put in a Russian camp, built in reprisal to a German concentration camp where thousands of Russian soldiers had perished. Later, when her identity was learned, she was taken to linger in a house in captured German territory under Russian surveillance. She suffered a slow death from a stomach ailment, lonely and remorseful, refusing to eat for fear of being poisoned."

Einstein said, "Goethe was right. 'The spirit I have summoned, I can no longer get rid of.' As I've said so many times before, 'No purpose is so high that unworthy methods in achieving it can be justified."

I felt sad. "Is there really no hope of improving the German mind? President Heuss told me last year that I should try to change your mind, if possible. He said that he would do anything in reparation for he still considers you the greatest German."

"I am a little astonished that he uses you, but no, I'm not," Einstein chuckled. "You are a Rhinelander-your delightful accent, your smile. But I'm not going to smile along. I have let Heuss know that because of the mass murder which the Germans inflicted upon the Jewish people, it is not possible for me to be associated in any way with any official German institution."

"A short time ago, I read the memoirs of Otto Abetz, the German Ambassador to Vichy France."

"Isn't he in jail, now?" asked Einstein.

'Yes," I replied, "he received twenty years after being caught in Germany by the French after the demise of the Thousand Years Reich. His scholarly mind and methodic thoroughness made his book an intellectual masterpiece, but a German masterpiece.

Scratch the skin of his intellect and what do you find beneath?"

Einstein chuckled, "I know too well. I worked with German intellects for many years. Of course, there are exceptions—von Laue, for one."

"And don't forget Lehmann-Russbuldt, and the thousands in the resistance movement."

Einstein interrupted bitterly, "But they were a drop of conscience in the ocean of apathy!"

"I wanted to admire Abetz for his historical honesty and accuracy, but his conscience was missing. The mysterious, which you told me was so important in man, was lacking. One sentence of his I'll never forget. He writes that he has fallen in disgrace through Hitler and Ribbentrop, but adds that this doesn't mean that they have fallen in disgrace through him. And true to form, he accepted dinner invitations by General von Stuelpnagel and other high military commanders in Paris who belonged to the conspiracy to assassinate Hitler, but never responded to their feelers on where he stood. He speaks in his book of the Janus face regarding leading French politicians, but what is with his face? He compares Hitler with Napoleon, saying that both great achievements were works of peace."

Einstein snapped, "Napoleon did not burn synagogues, but gave Europe a law book based on justice and equality for all."

"The saddest account in Abetz's book is the case of Grynszpan, who murdered von Rath, a member of the German Embassy. This Jewish teenager, he writes, wanted to revenge the extradition of his parents back to Poland. Abetz appeases his conscience by adding that the search for conspirators using the youth proved fruitless, and emphasizes again in another sentence that it could not be determined if the youth acted alone. One doesn't need to be a psychologist to see how he tries to justify himself. A few years before, I was in the German Embassy to receive the Cross of Honor for fighting on the front in the First World War, when his predecessor, Dr. Roland Koester, was forced by Gauleiter Wagner to enlarge the staff at the embassy by adding specially trained Nazi agents. Dr. Koester told me that I should escape to England to avoid being used by those agents for sinister purposes, with the promise of amnesty on returning to Germany. He confided in me that his life was also in danger, because he was appointed during the Weimar Republic and was still a democrat. He died late in a French hospital of a mysterious stomach ailment and it was whispered that he had been poisoned."

Einstein nodded, "This new ambassador knew how to sleep at night."

"It is for me inconceivable how the national mind of the German can take hold of the individual's conscience."

"Yes," said Einstein, "Nietzsche wrote once, 'When the memory says it happened, but the pride says no, the memory gives in.'"

Hitler used the Grynszpan murder as a pretext to collect from the Jews of occupied Europe many millions of gold marks as a ransom. I met a refugee in America who told me that Goering promised his family protection if he would give him a sketch by Duerer for his private collection."

Einstein grinned, "I know. After I left for America, my bank accounts and all my possessions were confiscated because the Germans supposedly found weapons in my house."

"The pathetic intellectualism of Abetz," I continued, "reminded me of my visiting German intellectuals after the Reichstag Fire. You said to me in 1930 that Germany could only be saved by changing the heart of man. I told them that I had witnessed the fire and am convinced that the fire was set by Goering, with the help of the Gestapo who used van der Lubbe, a wandering young Dutchman seeking work in Europe. Hitler had given himself away when he cried out, This is a cunning and well-prepared plot.' And Goering later raged at a reporter, grabbing his notebook, 'It was not one arsonist! It may be a good police report but not the kind of communique I have in mind! This was a signal for a communist uprising, and we shall arrest all of them.' In 1935 I talked with my cousin, Dr. Arthur Wolff, a noted lawyer of Duesseldorf, who observed the trial of van der Lubbe in Leipzig. His friend, who defended him, said that van der Lubbe was drugged throughout die trial and was often unable to speak. My cousin was forced to flee because he had defended communists accused of being part of the conspiracy.

"The brother of von Rath told me some years later in Wiesbaden that he felt Grynszpan was used by the Gestapo to kill his brother, who was against anti-Semitism. This did not surprise me since van der Lubbe was used for the Reichstag Fire and the Gestapo wanted to use me and some other twenty refugees, all former German front soldiers, for special purposes with the promise of receiving a new German passport to return home under the protection of the Fuehrer. I owe my life to Ambassador Koester who caused me to flee to England."

After a pause, Einstein said, "The greatest tragedy in my life was the discovery that in moments of decision involving conscience, scientists and religious leaders make a compromise with the state for reasons of security. Now I am old. So known in the world, yet so lonely."

I suddenly remembered my friends outside, and interrupted him, "Dr. Einstein! I forgot, I have guests waiting outside."

"There are guests waiting in the other room as well," said Miss Dukas, rising from the couch. I met Miss Dukas' reproachful eyes. She, of course, was Einstein's faithful guardian. She had to be discerning or he would be overwhelmed with visitors day and night. Yet, I had also an obligation toward a student, whom I had brought along. Einstein shall decide, I thought, and as I rose said, "Dr. Einstein, there is a boy, a physics student, who is terribly depressed, and if anyone can help him, you can."

"I?" His eyes opened wide. "Let him come in."

"His father is there, too," I added.

"Have them both come in; it's all right."

On my way out I said, "We must now speak English." Einstein just smiled, and when we three filed in, he stepped toward Bill with his hand extended in a warm greeting. After the brief introductions, I was eager to keep the conversation going and also involve Patrick.

"Dr. Einstein, since we last met, I have often wondered whether we could call energy some sort of life force, or vibration."

Einstein said, "There is no permanence in matter, but there is in energy. Matter combined with energy is the substance of the universe."

I took out my note pad and began jotting down key words.

The theory of relativity simplifies," he went on. "It does away with Newton's concept of absolute and independent space and time and, instead, stresses their unity. After many years of thinking about it, I had to discard Newton's gravitational force. My theory of gravitation postulates a curvature of the space-time unity that gives the motions of the planets a suitable background. The celestial substances — planets, nebulae, and rays of light—move according to geodesies. Within this curved space-time unity they follow the shortest possible route, the path of least resistance, from one point to another."

I looked at Pat and with my eyes urged him to say something. To my joy he asked, "Didn't you once call this the principle of least action? I remember reading about this for my class."

"Yes," answered Einstein, "by which I mean the principle of least time and least distance." He then mentioned the analogy of children playing marbles.

"There's a principle in man, too," I commented, "that moves him to obey his subconscious mind more often than his reason. It might be called the path of least mental resistance."

Einstein added, "Free will and reason will always capitulate when imagination and emotion are in control."

I felt ill at ease since we were still standing, so I made as if to offer Einstein a chair. But in the middle of my gesture, I suddenly realized the impropriety of my impulsiveness, all the more since we had experienced the same chair problems when I had visited him with the minister and my two nephews. Did Miss Dukas read my thoughts? Unexpectedly, she graciously invited us to sit down. I could hardly believe her change of mind, but Einstein obeyed, and so did we.

I felt more and more at home with him and spoke up confidently: "America's attitude toward Russia has split the academic community into two factions. To which do you belong?"

"You know I don't believe in the standardization of individuals," he answered, "and as for basic principles of world government, Russia's attitude is very foolish. But so is ours. I once proposed that the United States, Britain and Russia lay the groundwork for a supranational government, that they pledge their united military power as its guarantee. The small nations should then be invited to join. This system of government must have the right to intervene wherever the majority is ruled by a minority. I am still convinced that the best guarantee of freedom in the atomic age is a world federation ol States."

"Should this world government have the bomb?"

"We should have denounced atomic weapons even before we made proposals for controlling them internationally. This would have shown Russia that we agree unconditionally to international control, and that we don't intend to bargain, intimidate, or blackmail. The world government should have unlimited sovereignty, since it alone would know how best to defend humanity. Nations should have only limited sovereignty."

"But the Russians have refused, haven't they? They dismissed your suggestions."

"They ridiculed my idea of electing the deputees of the world government by free and independent elections in every nation. Then-distrust was justified, however, inasmuch as they asked whether Negroes enjoyed equal rights here."

"As far as I can see," interrupted Bill, "unfortunately they seem unwilling to yield their nationalism to supra-nationalism."

"I feel that at this time nationalism is a greater danger in America than in Russia," said Einstein. "Under the pretext of protecting the United States from Communism, the Committee on Un-American activities has resorted to witch-hunting, even against some distinguished generals of the last war. This fear of Communism seems to me a psychological camouflage. It is the way Hitler pointed to the danger from without, in order to secure his power from within. I am an opponent of every totalitarian system, whether Communistic or Christian. I believe in the versatility of the human mind and stand for its free development. And this is only possible when it is not bound to an institution, and when man relies on the regulative power of reason."

Bill asked, "And when man has no reason?"

"He nonetheless has a conscience," I countered.

"Conscience...." Einstein moved his head from side to side deliberatively. "Hitler gave us Jews the best compliment when he said that conscience is a Jewish invention. Unfortunately, conscience can be perverted and manipulated."

I said, "It exists, but is often overwhelmed by group thinking. Group living teaches us what is acceptable and what is not; man thus develops a sense of right and wrong according to the dictates of the group to which he belongs. This sense sometimes agrees with conscience, but often does not."

Bill suggested that the Nazi S.S. proved that conscience could be silenced, since they were able to teach young men to kill like automatons and feel no scruples.

"How true," I added, "at the conclusion of the bloody Roehm purge in which hundreds of people were brutally slain overnight, Hitler was able to announce without any public protest, Today I was the conscience of the German people."

"In that case," Einstein quickly added, "their Nazi pseudo-consciences would have troubled them if they hadn't killed."

Einstein's eyes sparkled while he talked and several times he looked over at Pat, as if encouraging him to partake in the conversation.

Glancing at my agenda, I saw the words, "Libel — Sheen" not yet checked off. "This American witch-hunting," I said, "was also directed toward you."

"Yes," Einstein replied, scratching his head. "Because I protested hearings before those Congressional committees which attempted to color people red, they call me a sympathizer of communist front organizations."

"But didn't you once admire Russia?" I asked.

"I have never admired any system that encourages a herd nature in man by suppressing his free will to choose for himself."

"I once read that you spoke well of Marx and Lenin."

"I said that Marx sacrificed himself for the ideal of social justice, but I didn't say that his theories are right. And Lenin" —he thought for a moment —"I don't believe he liked me." He suddenly looked angry. "How can I be called a communist when I have fought so long for freedom of thought, of expression, freedom from the military boot, and freedom from automation?"

"Dr. Einstein," I said, "you know it is rumored that you are an atheist...."

Drawing on his pipe, Einstein looked away, then smiled. "Haven't we talked about this before?"

"You said that one could call God the Urgeselz, or law of laws."

"Why not? You know you have a perfect right to name God any power you believe in. But what are you telling me when you say this? What's on your mind?"

"Professor, you know that in a Christian country, which America claims to be, it is dangerous for an educator to advocate a 'cosmic' religion. Think of the fate of Bertrand Russell, who scoffed at the moral precepts taught by Christian churches." Einstein smiled.

"This is no joking matter, Dr. Einstein; the libelers are busy. You know what Hitler wrote in *Mein Kampf*: 'Small lies are told every day, but when one tells big lies, the masses can't believe that the human mind could be so depraved as to invent them, and that there thus must be some truth in them.1" And pulling some papers from my pocket, "I have collected some statements about you."

Miss Dukas interrupted, "You have probably read what Dr. Einstein has already said about those slanderers: 'Arrows of hate have been shot at me, too. But they have never hit me, because somehow they belong to another world with which I have no connection whatsoever.'"

"Even though you may not be interested in this slander," I protested, "the world is. I have met many students, to say nothing of Americans at large, who have distorted ideas about you. I showed these statements to Bishop Sheen. As you know, he appears on television every week before millions of people."

"Yes," nodded Einstein, his eyes twinkling, "the Bishop is a dangerous converter and doesn't care for my cosmic religion."

"Bishop Sheen has agreed to rebuke anti-Semitic accusations and answer your attackers on television. He would have written personally, but felt that since a letter wasn't solicited, it might later be misinterpreted." Einstein laughed and tapped out his pipe in an ashtray.

"It's the tragedy of a minority," I continued, "to be persecuted for the actions of one of its members whenever those in power deem it necessary to substitute circuses for bread. You certainly haven't forgotten how Hitler, by putting the torch to the synagogues, transformed all of Germany into an arena of fire that went down in history as the Kristallnacht. Couldn't this or a book-burning happen here?"

"Not because of me!" flashed Einstein.

"You remind me of the lamb who wasn't important to himself, and so thought himself unimportant to the wolf." I leaned toward him. "Your books were burned in Berlin. I was there. And a student dropped a huge leather-bound volume, an illustrated Old Testament, and I stooped to pick it up. 'Leave that alone,' he shouted. It's a translation from the Hebrew.'"

Einstein smiled bitterly. "Of course, it wasn't a German book. One reason we Jews are so hated is because we gave the world the Bible."

"For the good of America," I persisted, "I need precise statements on God and Russia."

"Hah! About God, I cannot accept any concept based on the authority of the Church." Einstein's eyes smoldered with dark fire. "As long as I can remember, I have resented mass indoctrination. I do not believe in the fear of life, in the fear of death, in blind faith. I cannot prove to you that there is no personal God, but if I were to speak of him, I would be a liar. I do not believe in the God of theology who rewards good and punishes evil. My God created laws that take care of that. His universe is not ruled by wishful thinking, but by immutable laws."

"You know," I said, "I've wondered about this God myself, especially when I think of how many of the German generals and diplomats, like von Rundstedt and von Papen, were regular churchgoers and yet served Hitler."

"Being a churchgoer doesn't mean having moral concepts," snapped Einstein. Then, pulling on a tuft of hair, "We must show the Russians that the democratic way of life is based on law and morality, and that we are no threat, therefore, to her free development. But we must also show them that we're not afraid of them; we must convince them that our survival is dependent on theirs and theirs on ours. If they want to save themselves from an atomic war, they have to help create a world government."

"Didn't you once want to establish a court of wisdom?" asked Bill.

"Yes," and he explained that it had been discussed at the Harvard Tercentenary Conference. "It would have been a center for the propagation of ethics and morals, like the Sorbonne in medieval times. Such a seat of wisdom would represent the conscience of mankind and serve as a forerunner to the world government."

"But doesn't the concept of the State get in the way?" I asked.

"It's up to the individual to change that. No State has the right to dictate a man's conscience."

"Suppose, Dr. Einstein," I said after a short pause, "that Russia isn't ready?"

"Then we have to organize our world parliament alone, and Russia or China can't help falling in line sooner or later. I believe that most nations will keep their reason and good sense and become members of a world government rather than of a Communist bloc."

"Couldn't the United Nations function as a world government?"

"I'm afraid not," said Einstein. "Each nation there is free to arm and have atomic weapons. But it is most important to change our thinking, to change the heart of man. We must create a cosmic man, a man ruled by his conscience. You met your conscience at Verdun; you are the man we need! Bargaining won't do it; everyone of us must have a pure mind and pure intentions. It's foolish to think that in the long run one nation can hide its research from the others. It is the highest wisdom to share scientific discoveries with all. There is no alternative to a world government and international control of the atomic bomb; otherwise we'll all be destroyed."

"But what if our politicians don't understand how important this is and still think in terms of national interests and weapon races?" I questioned. "I speak as a sociologist, and one of the most important statements my teacher, Franz Oppenhcimer, ever made was this: 'Any group allied with the State to increase its political power is also allied with the army.'"

The Nazis would never have come to power had not Hitler allied himself with the Officer Corps: He needed the generals for his wars. He allowed the Kaiser and the aristocratic landowners to keep their estates and even squashed a scandal regarding the gift of an estate to President von Hindenburg. Hitler also allied himself with the industrial barons—Krupp, Stinnes, and Thyssen—assuring them lucrative armament business. Now this group alliance of politicians in Washington spells the same danger. So much of industry is involved in so-called defense work. It's even difficult for an employee to change jobs, since he faces a loss in various benefit plans."

Einstein nodded. There are too many private corporations without commensurate responsibility. If the old generation can't change its thinking or its values, then we must appeal to the younger generation, the politicians of tomorrow, and help them build a world government.

"Dr. Hermanns, as a survivor of Verdun, you must remind the youth that it is they who will lay their bodies on the Battlefields."

"Dr. Einstein, that reminds me of Fulton Sheen. You said he was a dangerous converter. He has helped me to become more

Jewish by studying Catholicism. As you know, I'm also a Quaker, a Protestant and a Yogi, because if I'm to found a cosmic man, thanks to your inspiration, then I must also know something about cosmic religion. Let me tell you what Bishop Sheen told me some years ago in Washington: The Church never fits in with the world, any more than Israel does, because they both have a divine calling to be a people apart and separated."'

"Still," said Einstein, "the Pope made an alliance with the Anti-Christ."

"I know. You do not have your gray hair for nothing. You are a prophet; you predicted this long before it happened." Einstein brushed aside the comment with his hand. He did not want to hear it. "It's a shame—a pact with God and Satan at the same time."

I said, "Somewhere I read that Napoleon once said, The Pope only needs to nod yes, and he dominates the consciences of a hundred million people.' But the tragedy is that Hindenburg had done no better, as Bruening told me some years ago in Harvard. The President eliminated him from his post of Reichskanzler unconstitutionally, without consulting the Parliament. He chose Papen as the successor. Although Catholic, he was at least a member of Hindenburg^ own caste. I reminded Bruening that Nietzsche once said; The German knows well how to slink to chaos.'—and: The deep, icy mistrust the German still arouses now whenever he gets into a position of power is an echo of that inextinguishable horror with which Europe observed for centuries that raging of the blond Germanic beast.' But what shall I do, Dr. Einstein? I still write German poetry. It is the language of my mother, and I feel we must help the Germans."

"Keep your goal of changing their hearts," Einstein said forcefully. "Create a community which develops the highest of man's qualities based on conscience. You must warn people not to make their intellect their god. The intellect knows methods but it seldom knows values, and they come from feeling. If one doesn't play a part in the creative whole, he is not worth being called human. He has betrayed his true purpose."

"I have brought a poem for you, Dr. Einstein. What you have

told me about intuition in all my conversations with you has inspired me":

FAREWELL INTELLECT

O dreams of power.
O power of dreams!
You pull and tear
man's fleshly seams,
recreate him with
atomic ribs,
soar up to visit
the Apocalypse.

O soul-filled love,
O love-filled soul!
Where will you end?
In the black hole
a trap, a turnstile
to oblivion? More dead
than death? A road-sign warned —
"Armageddon" it read.

There was one youth
shocked by the sign —
cried: "Greed farewell,
no longer mine.
Farewell scheming intellect —
my soul be my only property,
and intuition be my profit
to set my neighbor free. "

He heard a voice: "You will not be slain.
My son was lost and is found again. "

There was another moment of silence. I began putting my notes together, and Miss Dukas breathed a sigh of relief and asked with a smile whether we would like a quick cup of tea. This was, of course, understood as a goodbye gesture, and Bill politely refused with the excuse that we wanted to get back before dark. Miss Dukas rose, but seeing me scanning my memoranda again, exited through the curtain, where several guests were still at tea.

How could I have taken her hint? In large letters I had written the single word, "Patrick." He had been sitting quietly all this time. I quickly told Einstein in German that this young man was disillusioned with his studies and with the world in general and that he was close to suicide. Einstein looked over at him and smiled. "Pat," I urged, "didn't you want to ask Dr. Einstein something?"

"Dr. Einstein," Pat began slowly, "is there anything worth believing?"

Einstein bent forward. "Certainly there are things worth believing. I believe in the brotherhood of man and in personal originality. But if you asked me to prove what I believe, I couldn't. You can spend your whole life trying to prove what you believe; you may hunt for reasons, but it will all be in vain. Yet our beliefs are like our existence; they are facts. If you don't yet know what to believe in, then try to learn what you feel and desire."

As they talked, I was struck by the contrast between the two. Pat was a typical Harvard student. He wore casual slacks and a sport jacket, his tie was meticulously knotted and his full, bushy hair well-groomed. But Einstein, though almost sixty years his senior, was not of any age. Even the clothes he wore seemed timeless. The old creaseless trousers, the faded blue sweater, the sandals: I had seen them all a few years back at the Institute for Advanced Studies. His neck still projected boldly from his open collar as it had some twenty-five years ago in his pleasantly furnished apartment in Berlin. Even then it clearly announced: "I don't need to be adorned with a tie." And now it was topped by a white-haired head that seemed to belong to a patriarch over whom the centuries might roll, but who could never wither away. Our eighteen year-old student told me afterward that when Einstein spoke to him, he sensed the presence of eternity.

Still, Pat was very much in the present when he asked, "Does experience give us truth?"

"This is a difficult question," answered Einstein in a warm and fatherly manner. "One must not think in terms of truth when one sees objects, for truth is a concept that exists only in thought, and is expressed in sentences because sentences deal with concepts. You must first make contact with reality, then form an intelligent picture of it by building a theory around it. First you observe, then you give substance to what you observe by means of mathematics. In this sense, what is real becomes true."

"Dr. Einstein once said," I added, "that truth is a beautiful statue set in a desert. Winds try to bury it in sand, and man must constantly dig this sand away."

"It's up to each of us," said Einstein, "to also acquire knowledge independently of experience. If you think in pure thoughts, then you can comprehend reality. Experience may help, but you are endowed with enough reason to extend physical reality without experience. Behind this visible solar system, are many invisible ones; but, you need not go there in order to know this."

"Isn't truth inherent in man?" I interjected. "You once told me that progress is made only by intuition, and not by the accumulation of knowledge."

"It's not as simple as that," replied Einstein. "Knowledge is necessary, too. An intuitive child couldn't accomplish anything without some knowledge. There will come a point in everyone's life, however, where only intuition can make the leap ahead, without ever knowing precisely how. One can never know why, but one must accept intuition as a fact."

Suddenly Bill said, "This young man has always been brilliant in the sciences, but now he simply doesn't see why he should strive to achieve, whether for the immediate purpose of getting good grades or for long-term goals. He doesn't see why, in fact, life is worth living." Einstein turned back to Pat in some amazement.

"Dr. Einstein," I said, "his problem is the problem of youth all over the world. They have lost faith in what we, the older generation, have told them."

Einstein smiled. "Why shouldn't they doubt? I doubted everything they told me. I wanted to find out for myself." Then, turning to Pat, he asked, "Doesn't the question of the atomic and undulatory properties of light arouse your curiosity?"

"Yes, very much."

"Then what reason is there to ask, Why am I a physicist?' Isn't that enough to occupy your whole mind for a lifetime? Rely on your intuition. Many ideas will occur, of course, but examine each of them critically. You have liberty of choice, but don't use it like a writer of fiction. Try, rather, to think you're solving a crossword puzzle, and only one irreplaceable piece will fit. There are, for example, different systems of geometry to choose from, but there is only one equation which will fit your own sense experience."

"How will I know when I'm on the right path in my thinking?" Pat asked.

Einstein had settled back in his chair, but now leaned forward abruptly. "Don't think about why you question, simply don't stop questioning. Don't worry about what you can't answer, and don't try to explain what you can't know. Curiosity is its own reason. Aren't you in awe when you contemplate the mysteries of eternity, of life, of the marvelous structure behind reality? And this is the miracle of the human mind—to use its constructions, concepts, and formulas as tools to explain what man sees, feels and touches. Try to comprehend a little more each day. Have holy curiosity." I had to break in:

"Bertrand Russell has written about this." I then found the quote in one of my note pads:

It is difficult to turn Einstein's method into a set of textbook maxims for the guidance of students. The recipe would have read as follows: "First acquire a transcendent genius and an all embracing imagination, then learn your subject, and then wait for illumination. °

"Yes," added Einstein, "the only way to escape the personal corruption of praise is to go on working. One is tempted to stop and listen to it. The only thing is to turn away and go on working. Work. There is nothing else."

While Einstein was talking to Pat with such love and concern, I recalled a story about a family in Einstein's neighborhood. Every day after lunch, a little girl would take her notebook and disappear, and only say that she went to see a man who helped with her arithmetic. Concerned, her mother followed her one day, and saw her walk to the park and over to a bench, where she sat down and opened her book for the man to see.

The mother didn't believe her eyes. "Dr. Einstein, my daughter comes and bothers you every day with her schoolwork?"

"Oh, I don't mind," said Einstein. "Besides, she always brings me a cookie."

When Einstein stopped talking I asked, "Didn't you once tell a student to write a book on the theory of knowledge?"

"Yes, and if I had the time I would do it myself," he replied. "I wanted to write the history of scientific concepts in relation to sensory knowledge. It's one thing to observe experience by means of concepts, and another to apply concepts in ordering experience. I often tell students that concepts relate to sense-experience, not like the soup-to-the-beef but, rather, like the cloakroom-number-to-the-coat."

Pat interrupted, "How can I get rid of doubts about myself and about everything around me?"

Lowering his thick eyebrows, Einstein responded in a slow, deliberate manner. "First, you must have faith in an eternal world independent of you; then you must have faith in your ability to perceive it, and finally you must try to explain it by means of concepts or mathematical constructions. But don't always accept traditional concepts without reexamining them." Einstein laughed. "Even overthrow my relativity theory, if you find a better one."

"You once sent a friend a photograph, Dr. Einstein," I said, "with the dedication: Truth was at our fingertips, but vanished like a beautiful illusion.' What disillusioned you so much —perhaps mankind?"

"No," smiled Einstein. "My goal was to unify physics. For over twenty years I tried to build the electro-dynamic and quantum theories into my relativity theory, but I haven't succeeded." He turned to the young physics student. "You must believe that the world was created as a unified whole which is comprehensible to man. Of course, it's going to take an infinitely long time to investigate this unified creation. But to me that is the highest and most sacred duty—unifying physics. Simplicity is the criterion of the universe."

"What would you suggest?" asked Pat.

"Our aim," Einstein explained, "is to develop fewer and fewer concepts to explain natural phenomena, and to find a logical connection between them. Water was the primary substance for the Greek Thales. Later the assemblage of molecules was reduced to solid, liquid, and gaseous states. Science now deals with a tertiary system which tries to unify our sense impressions of the first and secondary systems. I have not yet mastered mathematics; I am mounting, mounting, mounting to reach the top. I've tried twenty times and more to put all the phenomena of nature into one formula. I used all the mathematical systems, but they would not work for me."

I remarked to Dr. Einstein, "Doesn't that affect your health? They said that you haven't felt well for a long time."

Einstein beamed at me, and then at all of us, one after another. He asked, "Do I look like someone who is no longer all here?" Turning to me, he said, "You should know that nothing bothers me." But after a moment he added, "No, something is disturbing me, and we owe this to science. Many of my colleagues don't see what is coming upon us, not just in America, but in the whole world: The mass destruction of humanity."

Pat looked at me as if to give me a cue and I said to Dr. Einstein, "When you were in Germany, you said that scientists should give their knowledge to mobilize the forces of nature for the benefit of man, not his destruction. Why is it that you stand so alone in your plan to create a supranational government?"

"Because," he replied, "men don't want to change their hearts. At the bottom of all problems stands the human animal, with his greed. Nothing is so dangerous for a nation as to try to obtain military supremacy. Yet, this is the plan, both here and in Russia. I've said this many times before: the alliance between industry and the military will destroy us, especially when combined with the attempt to militarize the young by ignorant politicians who have forgotten that 'he who lives by the sword shall die by the sword.'" Looking at me he continued, "Here is a survivor of Verdun, who can testify about a Battle where several hundred thousand young men perished in a few months."

I then interrupted, "The total at Verdun reached three-quarters of a million killed and at least as many wounded."

Einstein nodded, "Of course, no one wants to listen to me, yet as sure as I'm sitting here we will be destroyed unless we create a cosmic conscience. And we have to begin to do that on an individual level, with the youth that are the politicians of tomorrow. The Kaiser had tried," he said, turning toward me again, "to take over the conscience of his soldiers and we remember that Hitler said some years later that he would be the conscience of the German people. But no one, and certainly no state, can take over the responsibility that the individual has to his conscience. Yes, I am a lonely man with these views."

I asked him about Germany today: "President Heuss was shocked, for he told me about how he had written you a long letter asking you to come back to Germany. He promised to try to make good all that you suffered. But you responded with just a few lines:

Does the president think that I could forget the murder of six million Jews? To say that this was the work of the S.S. means nothing to me. They were Germans weren't they? History is not made by one man, hut by the masses.

Einstein responded, "For me, any man —and this applies especially to the Germans —who regards his own life as more important than that of his fellow creatures, is disqualified for life. I have
always had an inherent sense of justice and responsibility." He
chuckled, "That is why I was so disliked, I believe, by the Germans. I distrust the state that makes men in its image and likeness.
I despise men who love to march in equal step to the strains of a
music band. Nothing is so contemptible to me as heroism by order.
Look at this youth — and this is called Christian culture!"

He turned toward me, his white hair and large brown eyes
complimenting his serenity, despite the emotional subject: "If you
must tell Bishop Sheen something about mc, then tell him that I do
not lie, that I am an honest man."

After a pause, "Mr. Hermanns, do something with young people. Don't give up the goal to change the heart of man. The goal
may be beyond your powers, but it is rewarding. You know how it
is in life: first you're idolized, then stepped upon, and finally
cursed-keep your Rheinish humor. Be a loner. That gives you time
to write good poetry."

Einstein looked at Pat. "Be a loner. That gives you time to
wonder, to search for the truth. Have holy curiosity. Make your
life worth living."

Einstein then looked at me, "Let us be strong for truth and
make the intellect a servant of conscience. We Jews have one word
in Hebrew, 'mispat,' which means both love and justice. Too many
religions preach love without justice. As long as there are no decent conditions for all men, as long as religions do not consider
that their highest obligation is to create decent conditions for all
mankind regardless of what creed, they have degraded love and
justice."

"I forgot to tell you, Dr. Einstein, that the last conversation
with the minister provoked discussion in Church circles."

Einstein, smiling, placed his hand on his stomach, "I hope they
haven't given me this pain, too."

"Youth looks forward to reading my conversations with you.
And I shall dedicate them to youth."

"Your poems have more to say than I. Your poem on Verdun made a great impression on me."

"I would like to help youth by setting you as an example."

"There are people far worthier than me to be set as an example for youth."

"If you mean," I said, "people who have changed hearts, I have not forgotten Elsa Brandstroem, who saved hundreds of thousands of prisoners of war in Russia, and this as a citizen of a small neutral country, Sweden."

"Yes," Einstein energetically tapped the table, "and also that wonderful country, which has renounced war, gave the world Raoul Wallenberg. I nominated him for the Nobel Peace Prize in 1949 when I heard from survivors that he had saved some hundred thousand Jews in Hungary from Eichmann's gas chambers by giving them Swedish passports. Because he chose to help the Jews to the last, he could not escape the Soviet troops. He was captured and sentenced as a Nazi spy to a Siberian jail, where he is still said to be this day, in spite of official Swedish protests."

"Then I shall also dedicate my book to these two cosmic people, and my poem:

THE CHRIST

I gave you hands to open them to others,
  they are my hands.
I gave you feet to walk to your brothers,
  they are my feet.
I gave you a heart that knows of lonesome strife,
  it has my beat.
And as the shining sun gives breath of life
  to seas and lands,
so rise and walk and bless with hand and heart;
  you are of me and I of you a part.

Einstein drew his hand through his hair, "We must change the heart of man."

I saw the curtain move. Here was Pat, listening with head bowed, and Einstein accompanying his lecture with vivid electrifying gestures. The gray curtain, however, moved again. It alone seemed aware of time and space. I asked Einstein in German whether he would mind if Bill took pictures.

"Hah!" he smiled. "A souvenir again, I understand."

He escorted us to the porch. As Bill was setting up the camera, Pat pointed to a tree. "Can I truthfully say that that is a tree and, if so, what it means?"

"This could all be a dream," replied Einstein, matter-of-factly. "You may not be seeing it at all. But you have to assume something. Be proud of being the mean between macrocosm and microcosm. Stand still and marvel. Try not to become a man of success, but a man of value. Look around at how people want to get more out of life than they put in. A man of value will give more than he receives. Be creative, but make sure that what you create is not a curse for mankind."

Bill, who was placing us in position for the camera, countered, "But wasn't your equation used the wrong way?"

"I'm often asked that," said Einstein. "As an American I felt I had to inform President Roosevelt of the possibility that Germany would develop the bomb first. Looking back, I would have thought twice before writing that letter. Now they talk about an armament

Photograph of Einstein and the author, Fourth Conversation. 1954.

control, but that will never abolish wars. The armament industry is too powerful. The proverb, *Si vis pacem para bellum* (If you want peace, prepare for war) has proven false, and we must change the conditions that lead to war. And yet children still run around playing with guns in this society."

"If we have a Third World War," asked Pat, will anyone survive?"

"A few. There would probably be enough to start the race all over again, but all organic life would be affected."

"But where would they live if the world were poisoned?" I asked.

Einstein shrugged his shoulders. "In caves, I guess. And if they start a fourth world war, they will fight it out with clubs."

Miss Dukas had remained discreetly behind the screen door, but now she stepped outside and warned us that the occupants of a car, which had stopped in the middle of the street, were taking moving pictures of us all. That was the end of the interview; Einstein dashed behind the screen door, and I saw the white glimmer of his long hair blowing around his neck. For a moment I imagined a mountain that extends from earth into regions which man can never reach. What little he knows of it is the small path he is able to climb.

We were all hungry, so Bill drove us to the Princeton Inn and was our host for a generous meal. I contributed a bottle of wine, and we clinked glasses and drank to our positive vibrations, to the screen door, to the curtain, and to the secretary whose heart was, after all, bigger than we thought.

Pat suddenly became gloomy. "What did Einstein say? We will have a third world war?"

"Probably," I answered. "And then a fourth world war fought with clubs. The Germans taught me in school that 'war is the continuance of peace by other means.' If your generation, Pat, does not create a world youth movement and introduce a new ethic, one which will overthrow the powerful alliance between the army and the industrial giants, then that third world war will soon be upon us. It is your duty to help change this traditional thinking which results in the ultimate diplomacy— war."

On the way home, we all sank deep into our own thoughts for a while. Suddenly Pat said, "There seems to be something prophetic about Einstein. 'Have holy curiosity.' I'll never forget that. You know, Professor Hermanns, I had the feeling it wasn't Einstein speaking, but rather the spirit in him that spoke to the spirit in me."

I replied, "Einstein is a prophet!"

The next day as I was working on the conversation, I thought of Patrick and wrote this poem for him and other young people who see no purpose to live:

DIALOGUE WITH DESPAIR

Patrick from Without:

This earth is but one curse.
Who wants as slave to live?
I'm drowning, and a straw
religion wants to give
by saying, "You are saved."
I'm robbed to the skin by greed.
The mouth that speaks the blessing
my bony fingers feed.
A slave at the machine,
the bomb I make is blessed.
What system will survive
and carry me to rest?

Einstein from Within:

The world, a part of you,
behold yourself with awe.
You operate creation,
befriend the Cosmic Law.

What your free will decides,
moves holiness to soar,

or heats your blood to use
creation as a whore.

Is your will ego-drunk?
What greed wants, it must storm.
Is it a child of conscience
and gathers hearts to warm?

Your will calls cosmic workers.
These spirits come to dwell
in every cell and breath:
Are they from heaven or hell?

POSTLUDE

Albert Einstein died on April 18, 1955. His wish was that his brain be donated to science and the rest of his body burned.

EINSTEIN

"When dead, then let my worthless body burn,
My ashes throw into the silent air;
On earth no hole shall give my name a place."

We rowed through darkness with your precious urn.
There was a splash, a bubble sobbed somewhere.
Yet, friend, creation has preserved your trace.

The trace leads up; your pillow is a star.
Your dream? You see a golden ladder rise
And youths ascending, white and black and brown,

Rung after rung, reach out and touch afar
A light, heart-formed, that gleams into the skies.
The pilgrimage is endless up and down.

And they are changed. They radiate and glow,
Filled with your conscience and your intuition,
They act now with dimensions infinite.

Their will and purpose crush what's base and low
And stop the atom from its hell-bound mission.
They carry in their heart a holy writ:

We are cosmic substance, co-creators, we,
Of all that was and is and what shall be.

And these, my friend, were your last words to me:
"Have always holy curiosity."

About a generation later I contacted Pat in the East, who assured me that he has kept his promise to have Holy Curiosity as his leitmotiv for living. He asked me if I kept my promise regarding the founding of the World Youth Parliament. I immediately began the process of incorporating a non-profit foundation and, the vibrations of this dedication to world peace had its synchronistic effect at that most meaningful of all places for me, Verdun.

Epilogue

Bishop Boillon of Verdun, knowing me for many years through my pilgrimages to the battlefield and the legacy I received from Einstein, invited me to address the first international peace conference of the World Union of Martyred Towns, Peace Towns, held in Verdun in October 1982. As I was introduced to the mayors of Lidice, Warsaw, Coventry and Volgograd, I stopped while holding the hand of the representative from Hiroshima. I had on my tongue to say, "Einstein told me that he wanted to have the bomb explode on an uninhabited island in the Pacific with the Japanese High Command invited to observe, but Truman's military advisors rejected the idea. When Einstein heard of the tragedy, he shut himself up for eight days in mourning, refusing to see anyone." It was as if Einstein's spirit whispered to me: "Have you forgotten that I told you once not to defend me." I left this Buddhist monk in silence while the face of another churchman came to my mind: Fulton Sheen.

As a young Monsignor he liked to follow the hierarchical trend of thought, "The Jews are the out-group," and made light of Einstein's thoughts on Cosmic Religion by saying that one only needed to cross out the "s" in "cosmic" and the truth would be evident. When Sheen, my friend of many years, read one of my conversations with Einstein, he had a change of mind and in repentance sent me to ask Einstein for permission to make a special broadcast in New York about him. Einstein refused, because he did not care if the Catholics through the mouth of their most popular orator changed their opinion about him. Bringing the "No" back to Bishop Sheen, I said that I now understood why he, who sometimes dared to change his mind in opposition to the prevailing opinion of his superiors, had never become a cardinal, and if he had lived some centuries before might well have joined the long procession to the chamber of the Inquisition, in good company with Galileo and others.

Recounting these events in Stanford I was told by students that if Einstein had been a Catholic, Rome would acknowledge his saintliness. I answered, "Einstein's ashes had, as he wanted it, long been carried away with the waters into the ocean of oblivion. And if I were asked to write some words for posterity of this man who told me that if he had to come back to earth he would like to become a shoemaker and not a scientist, I would write this letter to Youth":

Youth of the world, in fulfilling Einstein's plea to found a Cosmic Religion as well as a World Youth Parliament, remember that your conscience must be nourished with spiritual food fitting your personal equation. For almost seventy years I have repeatedly called my vow into remembrance, replenishing it with Francis of Assists words, "God, make me an instrument of your peace, " or with Paramahansa Yogananda's advice, "Watch your spirit, for it is the carpet on which you walk to the next place of existence." The Cosmic Religion, Einstein suggested, does not destroy religious values of traditional beliefs, but embraces them in accord with the words which Jesus learned from his mother, "Hear 0 Israel, the Eternal is our God, the Eternal is one," and in acceptance of the challenge to his disciples, "You can do even greater things than I because I go to my Father. "Over the portals of the assembly house of the Cosmic Man, I would like you youth to write these words in honor of Einstein: "Salvation is self-awareness, the path leading to the Innermost Self— the Supreme Law or God. "

When some years ago Queen Elizabeth, the Queen Mother, after reading my book on Verdun, invited me to a luncheon in her St. James Palace, attended by four gentlemen-in-waiting, one of the young officers asked me under what principle the Cosmic Religion and the World Parliament of Youth would function. I would like you, my readers, to ponder my answer: You are not here by accident. Mobilizing your free will to create a new heart in you by your conscience, you will unfold psychic energy whose serene vibrations will help others whom you meet to see themselves in their relationship to their ego. The true value of the human being is cosmic significance transforming the karmic connection with your past. Let your conscience always create the significance of the moment. Have Holy Curiosity, as Einstein taught it.

Einstein visualized the composition of the Youth Parliament to include the Russians, Chinese and every nationality and race and to begin regardless of whether a nation rejects the invitation. Form study groups in the universities of every nation and wherever you go sow the vision of the Cosmic Man. Before you join any movement or group, examine its aims and beware those power structures which have as their motto, "We have the only truth." Learn to discern the difference between your conscience and the pseudo-conscience of a group. Never forget the power of your vibrations as a Cosmic Being. Truth is as infinite as the Cosmic Man, whose conscience is the voice of the Supreme.

When I, a young veteran of the Kaiser's war, visited Einstein in 1930, he pointed to the marching feet of the Hitler youth in the street below and asked, "When will the young people learn from history?" Youth of the World, my torch lit by my conscience on the bloodiest battlefield in history I pass on to you.

The Einstein-Hermanns Foundation is dedicated to the world family of the Cosmic Man. My readers are invited to help realize this goal by corresponding with me through the Foundation:

The Einstein-Hermanns Foundation P.O. Box 8129 Stanford CA 94305 U.S.A. Einstein, entrusting me with the mission of changing the heart of Man, inspired me to compose an anthem for the World Parliament of Youth, and I, in turn, entrust it to you, dear Reader:

UNITED HUMANITY

Unite Humanity, unite —
one fellowship on earth!
Too long we turned our back to peace,
too long life had no worth.
The sun, the earth and you are made
to share, to wonder, not to hate.
Draw we not hope with our first breath?
Hope knows no gun, no death.

Come brothers from the desert heat,

and from the icy shore, let us together build one world,
and marching feet no more.
One world, one peace where you with me
can meet as brothers proud and free.
Is not the sun, the earth, the air
for you and me to share?

Your child, my child, is it not born
and nursed in love to live;
not handed down from Mount Sinai
one law: Receive to give?
Come let us vow:
Your son, my son
no longer shall embrace the gun.
Behold, one light shines from above —
one dignity, one love!

Einstein And The Poet —In Search Of The Cosmic Man

by William Hermanns

Centering on the close 34-year relationship with Einstein, the author begins this absorbing book by describing his vow on the battlefield oi Verdun: "God, save me, and I will serve you as long as I live." A member of the League for Human Rights, the Alexander von Humboldt International Club, and other peace organizations, Professor Hermanns became a disciple of Albert Einstein, His four conversations with that genius is nothing less than a pilgrimage to the altar of world peace.

Professor Hermanns had his first conversation in 1930 in the very Hat of Einstein, An educational play-writer for the German Radio, he had hoped to use the radio to defend the democratic ideals of the Weimar Republic, and to show the German people that a genius is made from within and not through shouts of "HEIL" from the masses. While the two conversed, Nazi Youth marched to the slogan: "When Jewish blood spurts from our knives, then it will be twice as good." Outside Einstein's door, a man threatened to set oft' a bomb.

In their second conversation at Princeton in 1943, they discussed the nature and existence of matter, encounters with Max Liebermann, Gcrhart Hauptmann, Magnus Hirschfeld and, of special interest, the conspiracy of the Empress in the Crown Prince Palace. The author shares his experience in the demise of the Rathenau Society and freedom in the first years of the Nazi terror in Germany, and Einstein explains his change from an absolute to a conditional pacifist,

In the third conversation in 1948, they realized the need for the founding of a cosmic religion; they also discussed Jesus, Elsa Brandstroem, Planck. Haber, Nernst and the Nazis, the Church and anti-Semitism. In the fourth, shortly before Einstein's death, they talked about Bruening, Breitschied and Empress Hermina. The author also records the advice given by Einstein to a Harvard student bent on suicide and Einstein's dream of a World Youth Parliament to avert a third world war in which only a quarter of humanity would survive,

In this exceptional book, which contains documents published for the first time, the reader quickly understands that Professor Hermanns' mission is to accomplish Einstein's goal to found a cosmic religion based on one's innate conscience in order to prevent an apocalyptic third world war.

Made in the USA
Las Vegas, NV
01 November 2024

10807983R00098